Lecture Notes in Computer Science 13183

More information about this subseries at https://link.springer.com/bookseries/7412

Moi Hoon Yap · Bill Cassidy ·
Connah Kendrick (Eds.)

Diabetic Foot Ulcers Grand Challenge

Second Challenge, DFUC 2021
Held in Conjunction with MICCAI 2021
Strasbourg, France, September 27, 2021
Proceedings

Editors
Moi Hoon Yap
Manchester Metropolitan University
Manchester, UK

Bill Cassidy
Manchester Metropolitan University
Manchester, UK

Connah Kendrick
Manchester Metropolitan University
Manchester, UK

ISSN 0302-9743 ISSN 1611-3349 (electronic)
Lecture Notes in Computer Science
ISBN 978-3-030-94906-8 ISBN 978-3-030-94907-5 (eBook)
https://doi.org/10.1007/978-3-030-94907-5

LNCS Sublibrary: SL6 – Image Processing, Computer Vision, Pattern Recognition, and Graphics

This Springer imprint is published by the registered company Springer Nature Switzerland AG
The registered company address is: Gewerbestrasse 11, 6330 Cham, Switzerland

Preface

This is the first proceedings of Diabetic Foot Ulcer (DFU) research focusing on the DFU Challenge 2021 (DFUC 2021), organized in conjunction with the 24th International Conference on Medical Image Computing and Computer Assisted Intervention (MICCAI 2021). Due to the COVID-19 pandemic, the challenge event was conducted online on September 27, 2021. We received 500 submissions for the validation stage and 28 submissions for the testing stage.

The overall goal of the DFU challenge is to solicit original works in DFU and to promote interactions between researchers and interdisciplinary (national and international) collaborators. It aims to motivate the health care domain to share datasets, participate in ground truth annotation, and enable data-innovation in computer algorithm development. In the longer term, it will lead to improved patient care and reduce the strain on overburdened healthcare systems. With joint efforts from the lead scientists of the UK, US, India, and New Zealand, these inaugural challenges were well-received, with the first DFU challenge successfully conducted in 2020 for the task of DFU detection. The task of DFUC 2021 related to multi-class classification, for the purpose of supporting research towards more advanced methods of DFU pathology recognition.

This proceedings provides an overview of the development of DFU datasets, gathers methodological papers of classification methods evaluated at the DFUC 2021 along with a DFUC 2021 summary paper from the organizers, and considers post challenge papers. Apart from the Overview and Summary paper, all papers were reviewed by three reviewers and assigned to a meta-reviewer. For DFUC 2021 papers a single-blind review process was used, accepting only the top five entries due to the proceedings call for papers deadline. For the post challenge papers, a double-blind review process was completed with a 25% acceptance rate. In total we received four post challenge papers with only one accepted. The organizers were not listed as authors for the challenge and post challenge papers.

As a concluding note, the organizers of the challenge continue to support the research community by providing a live leaderboard to test the performance of their algorithms. To date, there are 350 submissions on the live leaderboard, with approximately 10 submissions per day. Researchers who are interested in the challenge can request the datasets and evaluate the task on our grand challenge websites (DFU detection task and DFU classification task). The organizers will conduct DFUC 2022 in conjunction with MICCAI 2022, with the task focusing on DFU segmentation.

December 2021

Moi Hoon Yap
Bill Cassidy
Connah Kendrick

Organization

General Chairs

Moi Hoon Yap	Manchester Metropolitan University, UK
Bill Cassidy	Manchester Metropolitan University, UK
Neil Reeves	Manchester Metropolitan University, UK

Organizing Committee

Moi Hoon Yap	Manchester Metropolitan University, UK
Neil Reeves	Manchester Metropolitan University, UK
Andrew Boulton	University of Manchester and Manchester Infirmary, UK
Satyan Rajbhandari	Lancashire Teaching Hospitals, UK
David Armstrong	University of Southern California, USA
Arun G. Maiya	Manipal College of Health Professions, India
Bijan Najafi	Baylor College of Medicine, USA
Bill Cassidy	Manchester Metropolitan University, UK
Justina Wu	Waikato District Health Board, New Zealand

Clinical Chairs

Joseph M. Pappachan	Lancashire Teaching Hospitals, UK
Claire O'Shea	Waikato District Health Board, New Zealand

Technical Chairs

Connah Kendrick	Manchester Metropolitan University, UK
David Gillespie	Manchester Metropolitan University, UK

Program Committee Chair

Connah Kendrick	Manchester Metropolitan University, UK

Program Committee

Christoph Friedrich (Area Chair)	University of Applied Sciences and Arts Dortmund, Germany
Azadeh Alavi (Area Chair)	Royal Melbourne Institute of Technology, Australia
Salman Ahmed	National University of Computer and Emerging Sciences, Pakistan
Nora Al-Garaawi	University of Kufa, Iraq
David Asher	Baker Heart and Diabetes Institute, Australia
Raphael Brüngel	University of Applied Sciences and Arts Dortmund, Germany
Bill Cassidy	Manchester Metropolitan University, UK
Adrian Galdran	Bournemouth University, UK
Manu Goyal	UT Southwestern Medical Center, USA
Orhun Güley	Technical University of Munich, Germany
Christian Igel	University of Copenhagen, Denmark
Abdul Qayyum	National Engineering School of Brest, France

Sponsors

Contents

Development of Diabetic Foot Ulcer Datasets: An Overview

Moi Hoon Yap[1(✉)], Connah Kendrick[1], Neil D. Reeves[4], Manu Goyal[3], Joseph M. Pappachan[2], and Bill Cassidy[1]

[1] Department of Computing and Mathematics, Manchester Metropolitan University, Manchester M1 5GD, UK
M.Yap@mmu.ac.uk
[2] Lancashire Teaching Hospitals NHS Trust, Preston PR2 9HT, UK
[3] Department of Radiology, UT Southwestern Medical Center, 5323 Harry Hines Blvd., Dallas, TX 75390-9085, USA
[4] Musculoskeletal Science and Sports Medicine, Manchester Metropolitan University, Manchester M1 5GD, UK

Abstract. This paper provides conceptual foundation and procedures used in the development of diabetic foot ulcer datasets over the past decade, with a timeline to demonstrate progress. We conduct a survey on data capturing methods for foot photographs, an overview of research in developing private and public datasets, the related computer vision tasks (detection, segmentation and classification), the diabetic foot ulcer challenges and the future direction of the development of the datasets. We report the distribution of dataset users by country and year. Our aim is to share the technical challenges that we encountered together with good practices in dataset development, and provide motivation for other researchers to participate in data sharing in this domain.

1 Introduction

The Diabetic Foot Ulcer (DFU) is one of the major complications resulting from diabetes, which can lead to lower limb amputation [1]. Regular foot check by clinical professionals is required for patients with DFU development, which is often costly with expensive medication and/or referral to specialist care [2]. Research shows that healthcare services that treat DFU are unable to handle the growing number of patients due to inadequately trained medical staff [3], which is especially prevalent in low-income countries and rural areas [4,5].

Over the past decade, the development of digital and information technology has enabled the creation of new computer-based solutions for healthcare, including wound care [6]. Figure 1 illustrates the timeline of development of DFU datasets, including the first use of computer vision methods in DFU detection. The focus of the earlier DFU research is in the design of new capturing tools [7], initiated in 2015. At the same time, computer vision methods (based on image processing algorithms) and conventional machine learning methods were used to analyse those images [8]. With the advent of deep learning in computer

M. H. Yap et al. (Eds.): DFUC 2021, LNCS 13183, pp. 1–18, 2022.
https://doi.org/10.1007/978-3-030-94907-5_1

vision tasks, researchers began investigating the use of deep learning for DFU segmentation in 2016, where the first fully automated segmentation paper was published in 2017 by Goyal et al. [9].

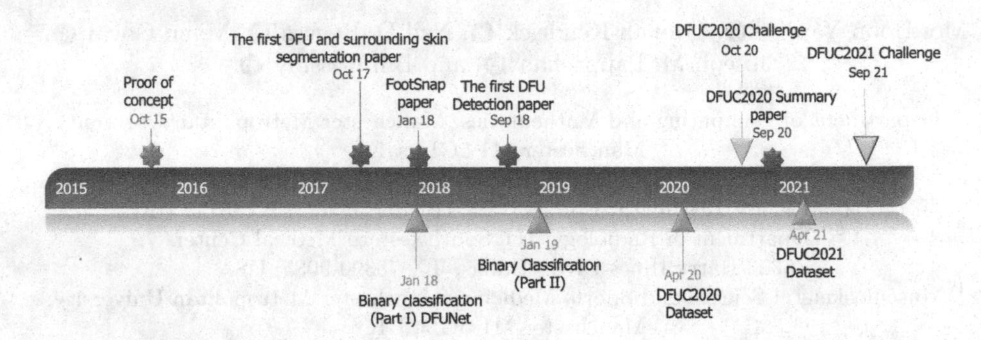

Fig. 1. The timeline of the development of DFU datasets and analysis. The proof-of-concept in using computer vision for DFU analysis was published in October 2015 by Yap et al. [8]. Although the FootSnap app was created in 2015, the paper was not published until 2018 [7].

Motivated by the success of deep learning on DFU segmentation, the team at the Manchester Metropolitan University and Lancashire Teaching Hospitals NHS Foundation Trust had obtained ethical approval from the UK National Health Service Research Ethics Committee (reference number: 15/NW/0539) to create larger scale imaging datasets of DFUs, with approval to share the datasets with the research community for research and research challenges, provided that the users abide by the licence agreement. The first dataset (Part I) was on binary classification using normal and DFU patches, where the authors designed a new deep learning network (DFUNet) and benchmarked the datasets with popular networks at that time [10]. The second dataset (Part II) was also on binary classification, and focussed on ischaemic and infection skin patches. The binary classification of ischaemia-vs-all and infection-vs-all were benchmarked by Goyal et al. [11], and they proposed an ensemble method to increase the accuracy of infection and ischaemia recognition. In 2020, the team conducted the first inaugural research challenge in DFU detection, DFUC2020 [12], and the second challenge in DFU multi-class classification in 2021 [13].

The remaining sections of the paper are organised as follow: Sect. 2 provides a survey of DFU data capturing methods; Sect. 3 reviews the available DFU datasets; Sect. 4 describes the DFU research challenges conducted over the past years; Sect. 5 presents future work and research directions of DFU analysis; and Sect. 6 summarises the paper.

2 A Survey of Data Capturing Methods

In current clinical practices, podiatrists and consultants use a range of digital single-lens reflex (SLR) camera models to collect DFU photographs [10,14,15]. The photographs are transferred to a secured storage which is often isolated from the patients' electronic health records. The process is operator dependent, with poor consistency across different clinic and care settings. Due to these inconsistencies and limitations of 2D images, it has not been possible to quantify the changes of the ulcers over time.

Over the past few years, several research teams have proposed new methods in standardising data capture of DFUs. The earliest attempts were conducted by Wang et al. [16] and Yap et al. [8]. Wang et al. [16] developed a smartphone app capable of image segmentation of DFU wounds using an accelerated mean-shift algorithm, which pre-dates current deep learning approaches. To improve and standardise the acquisition of DFU images, they created an image capture box, to be used in conjunction with the smartphone app. Their goal was to promote a more active role in patient self-monitoring. This approach used skin colour to determine foot boundaries, while wound area is determined by a simple connected region detection method. This system also assessed healing status using a red-yellow-black colour evaluation model and a quantitative trend analysis of time records for a given patient. The system was tested with 34 patients and 30 wound moulds. Figure 2 shows the design of the capture box[1].

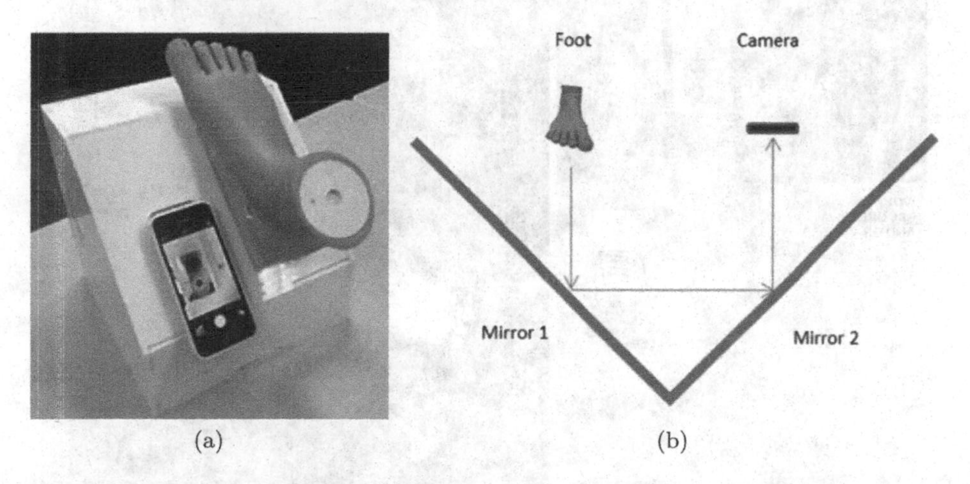

(a) (b)

Fig. 2. Illustration of the DFU image capture box proposed by Wang et al. [16]: (a) image capture box with smartphone and foot with wound mould attached, (b) capture box optical principle.

[1] Reproduced with permission from Peder C. Pedersen, Department of Electrical and Computer Engineering, Worcester Polytechnic Institute, Worcester, MA 01609, USA.

Wang et al. [17] later validated the capture box in a clinical study using 32 DFU photographs obtained from 12 diabetic patients. This system used a laptop to perform wound segmentation, calculation of the wound area and calculation of a healing score. They found a good correlation between human and automated wound area measurements, reporting a Matthews correlation coefficient value of 0.68 for the wound area determination algorithm, and a Krippendorff alpha coefficient within the range of 0.42 to 0.81.

Wang et al. [18] conducted a third study using their capture box, which collected 100 DFU photographs from 15 patients during a 2-year period. In this study, they utilised superpixel segmentation, using the Simple Linear Iterative Clustering (SLIC) algorithm, as inputs for a cascaded two-stage SVM-based (Support-Vector Machine) classifier to determine wound boundaries for DFU. Colour and bag-of-word representations of local dense scale invariant transformation features are used as descriptors for excluding non-wound regions. Wavelet-based features are used as descriptors to identify healthy tissue from wound tissue. Finally, wound boundaries are refined by applying a conditional random field method. This system ran on a Nexus 5 smartphone, and reported an average sensitivity of 73.3%, and an average specificity of 94.6% with a computation time of 15 to 20 s.

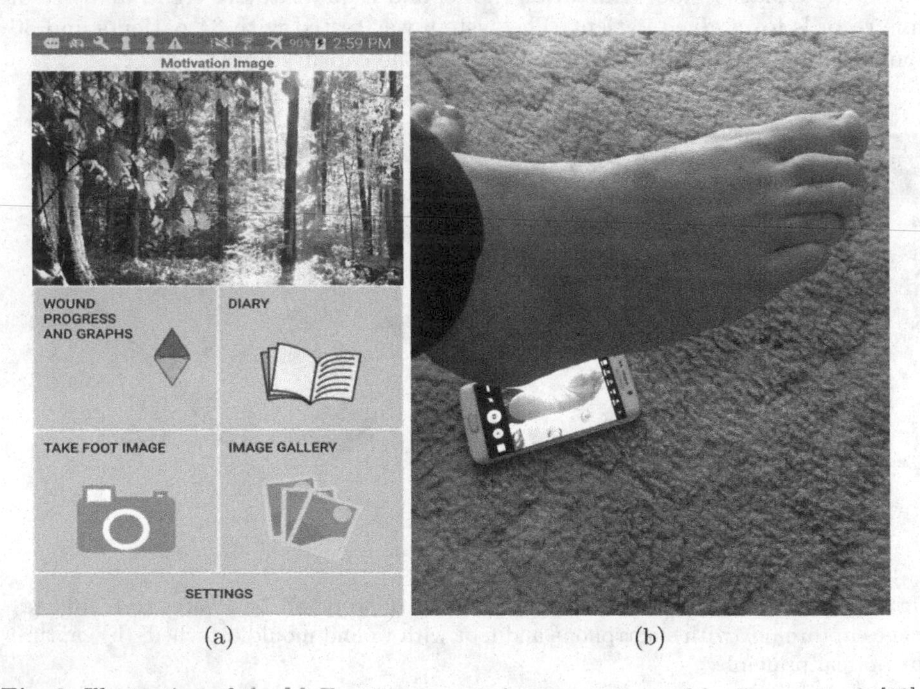

Fig. 3. Illustration of the MyFootCare smartphone app, created by Brown et al. [19]: (a) main screen showing all available features, (b) demonstration of the voice-guided photograph acquisition feature.

In 2015, Yap et al. [8] created a new mobile app called FootSnap which was used for standardising photographic capture of the plantar aspect of feet. This system generated an outline of a foot from an initial photograph of the patient's foot. The outline could then be recalled on-screen to help align the foot in subsequent photographs. The app was initially tested on healthy feet, then later evaluated for standardisation on clinical data [7]. The app was then used as part of a clinical trial investigating DFU prevention using smart insole technology published in 2019 [46].

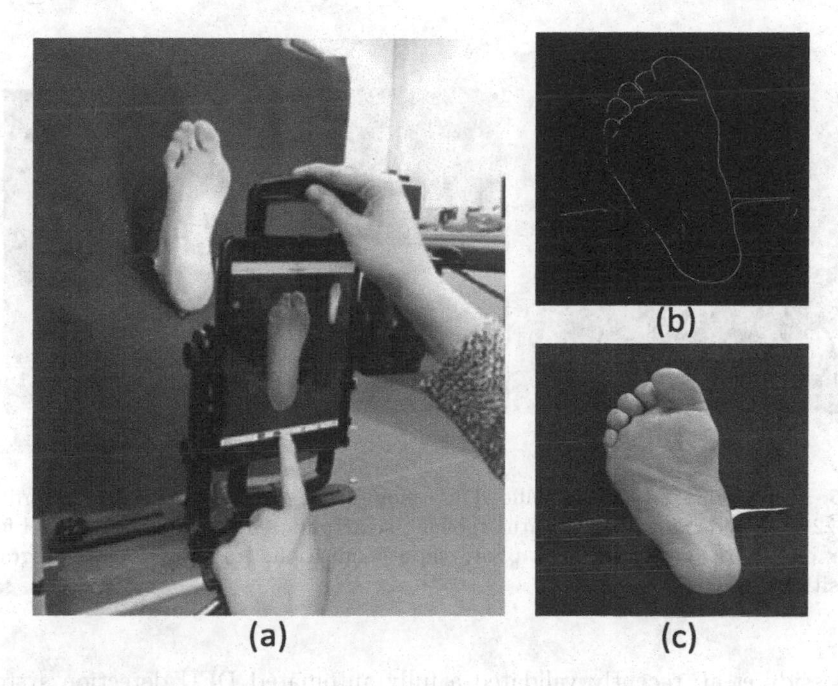

Fig. 4. Illustration of the FootSnap mobile app: (a) the data capturing tool and settings; (b) ghost image generated automatically to standardise data capturing; and (c) plantar aspect of foot captured.

Researchers would continue to investigate the use of mobile technologies to address the growing problem of DFU. Brown et al. [19] developed the MyFootCare app - a self-monitoring tool that could be used by patients to promote self-care in home settings. The app was evaluated with 3 DFU patients, who reported it as useful for tracking wound progress and to assist in communications with clinicians. The app implements a number of features, including image capture, wound analysis (using OpenCV), diary reminders, and wound size tracking using a graph. The image capture allows the patient to place the phone onto the floor with the screen facing upwards. The patient can then guide their foot into the correct position, with voice feedback provided by the app. This allows the patient to remain seated while positioning their foot for image capture. However, this functionality was not completed in time for the evaluation with patients. Figure 3 shows the design of the app and its proposed use in DFU photograph acquisition[2].

More recent techniques for photographic acquisition of DFU have been proposed by Swerdlow et al. [20]. They devised a "foot selfie" device comprising an elaborate assembly which helps to position the foot in front of a smartphone while minimising surface contact (see Fig. 5)[3]. Compared to some of the earlier capture methods, this solution has the advantage of not requiring contact between wound and surface, and also allows for capture of more than just plantar DFU.

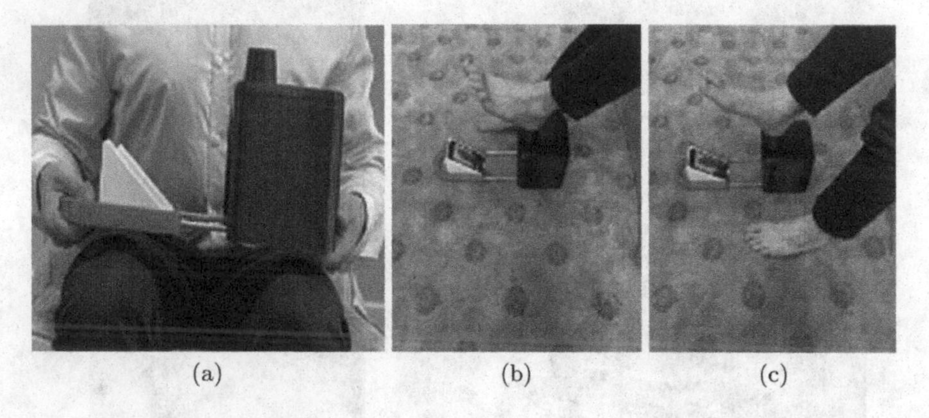

Fig. 5. Illustration of the "foot selfie" DFU monitoring system proposed by Swerdlow et al. [20]: (a) side view of apparatus showing smartphone holder on the left and foot holder on the right, (b) left foot photograph acquisition, (c) right foot photograph acquisition.

Cassidy et al. recently validated a fully automated DFU detection system which utilised mobile and cloud-based technologies [21,22]. This system used a cross-platform mobile app to capture photographs of patient's feet in clinical settings that could be uploaded to a cloud platform for inference to detect the presence of DFU. The system was validated in a proof-of-concept study at two UK hospitals over a 6-month period. This technology is currently being adapted for use in additional clinical studies with the aim of replacing SLR cameras in the acquisition of DFU photographs. Figure 6 shows a selection of screens from the cross-platform mobile app.

[2] Reproduced with permission from Ross Brown, School of Computer Science, Science and Engineering Faculty, Queensland University of Technology, Brisbane, Queensland, Australia.

[3] Reproduced with permission from David G. Armstrong, Southwestern Academic Limb Salvage Alliance (SALSA), Department of Surgery, Keck School of Medicine of University of Southern California, Los Angeles, California, USA.

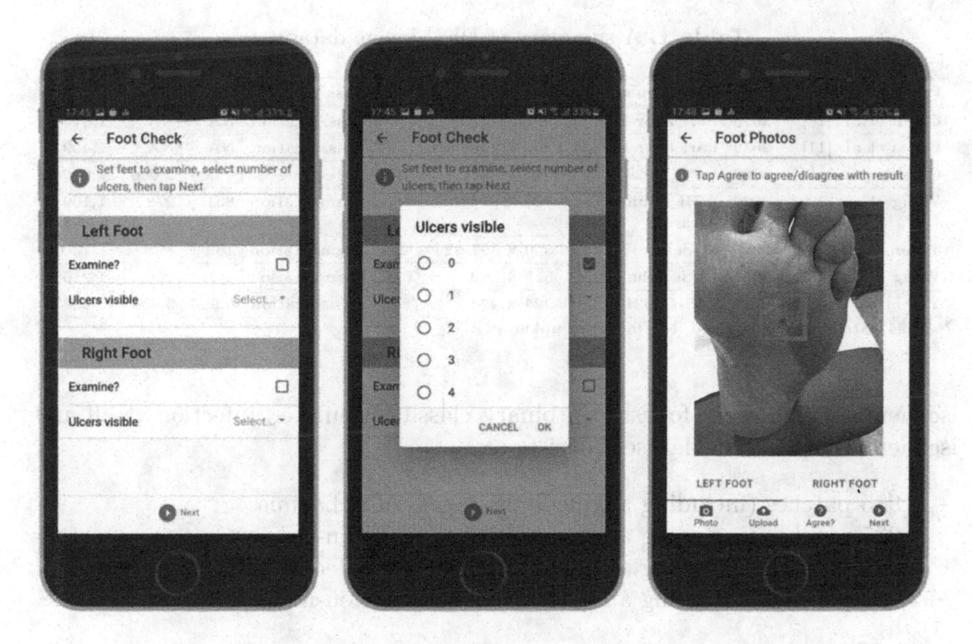

Fig. 6. Illustration of the cross-platform mobile app used to capture DFU photographs at two UK NHS hospitals during a six month proof-of-concept clinical evaluation.

3 A Review of DFU Image Datasets

This section reviews the available public DFU image datasets and the number of users (by countries, if known). Table 1 summarises and compares seven publicly available datasets.

3.1 Binary Classification

DFU Patches and Normal Patches (Part A or Part I). The Part A DFU Dataset [10] consists of 1038 DFU patches and 641 normal patches. This is the first binary classification dataset shared with the research community. The ground truth was produced by two healthcare professionals, specialising in DFU, using the annotation tool developed by Hewitt et al. [27]. The authors introduced this dataset and created a new deep learning model, DFUNet, to benchmark the performance on the dataset. DFUNet achieved an F1-score of 0.939. Since its initial release, this dataset has been used for research, with the most recent publication achieving the best performance of 0.952 in F1-score [28].

Recognition of Infection and Ischaemia Datasets (Part B or Part II). The first infection and ischaemia datasets were created by Goyal et al. [11], which consists of 1,459 DFUs: 645 with infection, 24 with ischaemia, 186 with infection and ischaemia, and 604 control DFU (presence of DFU, but without infection or

Table 1. A summary of DFU image datasets.

Publication	Year	Dataset name	Resolution	Task	Train	Test	Total
Goyal et al. [10]	2018	Part A or Part I	Varied	Classification	NA	NA	1,679
Goyal et al. [11]	2019	Part B or Part II	256 × 256	Classification	NA	NA	1,459
Cassidy et al. [15]	2020	DFUC2020	640 × 480	Detection	2,000	2,000	4,000
Wang et al. [23]	2020	AZH wound care dataset	224 × 224	Segmentation	831	278	1,109
Thomas [24]	NA	Medetec	560 × 391 224 × 224	Segmentation	152	8	160
Wang et al. [25]	2021	FUSeg challenge	512 × 512	Segmentation	1,010	200	1,210
Yap et al. [26]	2021	DFUC2021	224 × 224	Classification	5,955	5,734	15,683*

* 3.994 patches are unlabelled; NA indicates unknown

ischaemia). This work focused on binary classification, i.e., infection-vs-all and ischaemia-vs-all. This dataset consists of:

- 4,935 patches (including augmented images) of ischaemia
- 4,935 patches (including augmented images) of non-ischaemia
- 2,946 patches (including augmented images) of infection
- 2,946 patches (including augmented images) of non-infection

Image labelling information:

- 00XXXX_1X.jpg - the original image patch
- 00XXXX_2X.jpg and 00XXX_3X.jpg - natural data augmentation, where M indicates mirroring; R1, R2 and R4 indicate rotations.

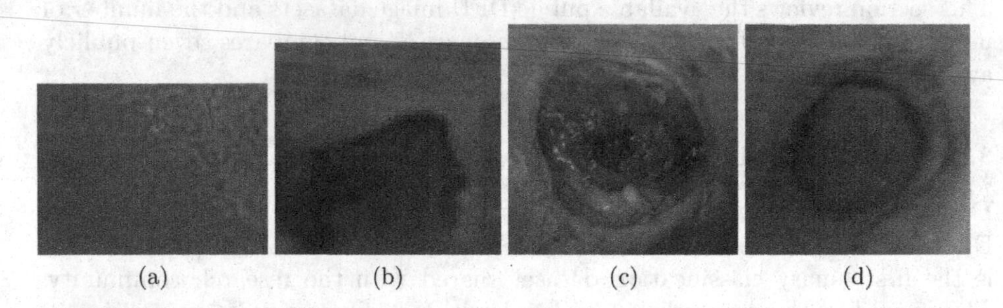

Fig. 7. Comparison of image patches from Part A and Part B datasets: (a) normal patch from Part A, (b) abnormal patch from Part A, (c) infection patch from Part B, and (d) other patch from Part B.

Goyal et al. [11] introduced an ensemble CNN method for binary classification and achieved an F1-score of 0.902 and 0.722 on ischaemia-vs-all and infection-vs-all respectively. Since the release of this dataset in early 2020, Al-Garaawi et al. [28] achieved the best F1-scores of 0.990 and 0.744 on ischaemia-vs-all and infection-vs-all, respectively. This demonstrates a significant challenge for machine learning algorithms in recognising infection from other classes.

Figure 7 compares the Part A and Part B datasets. It is noted that the Part A dataset did not include the whole ulcer region, as it consists of image patches cropped from examples exhibiting ulcers and non-ulcers. In contrast, the Part B dataset consists of ulcers with different pathologies.

3.2 DFU Detection (DFUC2020)

In 2018, Goyal et al. [29] reported the performance of object detection algorithms on an in-house DFU dataset with 1,775 images. They proposed the use of a 2-tier transfer learning method using Faster R-CNN with InceptionV2 model. Overall, they achieved the best mean average precision (mAP) of 91.8% on 5-fold cross-validation.

Cassidy et al. [15] introduced the largest DFU detection dataset to date, which was released on the 27th April 2020. This dataset contains 4,000 images (largely DFU images, with a small proportion of non-DFU images in the testing set), is highly heterogeneous and includes numerous challenging examples to help ensure robustness in algorithm development. Figure 8 shows an example of the expert labelling provided by podiatrists for this dataset.

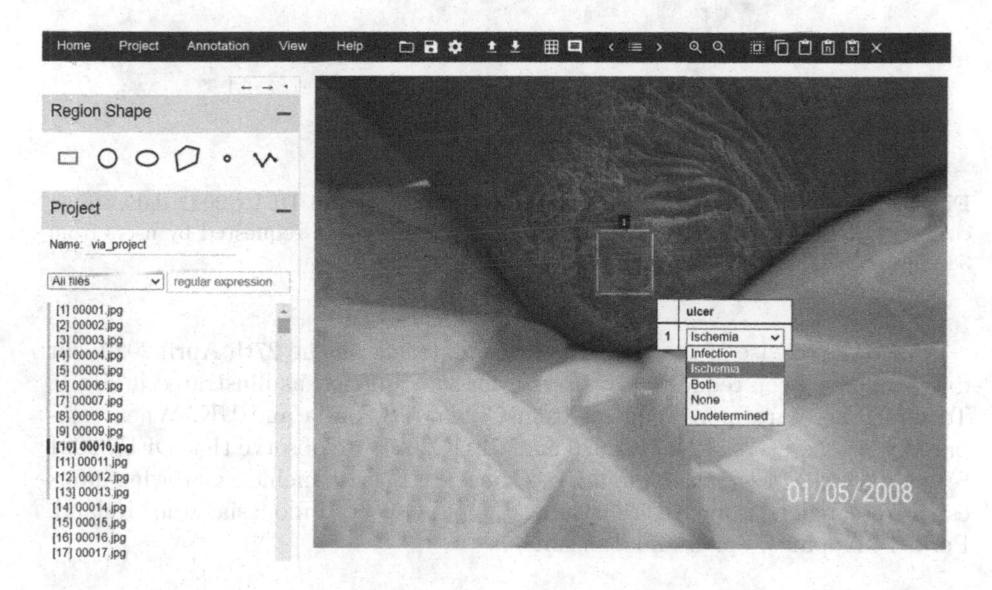

Fig. 8. Illustration of expert labelling provided for the DFUC2020 dataset using the VGG Image Annotator [30]. Patches from this dataset, together with class labels would later be used in the DFUC2021 dataset.

3.3 Multi-class DFU Classification (DFUC2021)

The first multi-class DFU classification dataset was released on the 27th April 2021 by Yap et al. [26], with four types of DFU patches, i.e., control, ischaemia, infection and both (co-occurrence of ischaemia and infection). This dataset is comprised of cropped ulcer regions from the DFUC2020 dataset [15] and the Part B dataset [11].

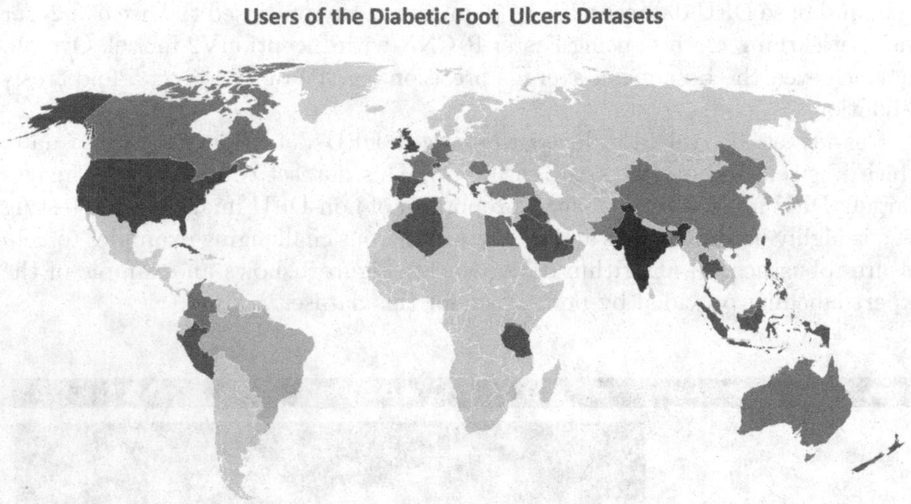

Users of the Diabetic Foot Ulcers Datasets

Fig. 9. Distribution of researchers using the DFUC2020 and DFUC2021 datasets by country. At the time of writing this paper, the datasets were requested by users from 38 countries. This map generation is powered by Bing.

Since the release of the DFU Challenge datasets on the 27th April 2020, our datasets have been requested by users from 38 countries, as illustrated in Fig. 9. To date, there are more requests from China, US, India and UK. When compared with the users of DFUC2020 and DFUC2021, we observe that DFUC2021 reached a wider research community as shown in Fig. 10, including growing interest of researchers from Algeria, Cuba, Egypt, Greece, Indonesia, Iraq, Ireland, Peru, Philippines, Spain and Tanzania.

3.4 Other Datasets

Following the successful release of the Part A, Part B and DFUC2020 datasets by researchers at Manchester Metropolitan University and Lancashire Teaching Hospitals [12,15,31], other research groups have followed this path and released their own public DFU datasets. Wang et al. [23] released the Foot Ulcer Segmentation Challenge dataset which focuses on single class semantic segmentation of foot ulcer wounds and contains 1,210 images with 1,010 labels, of which 200

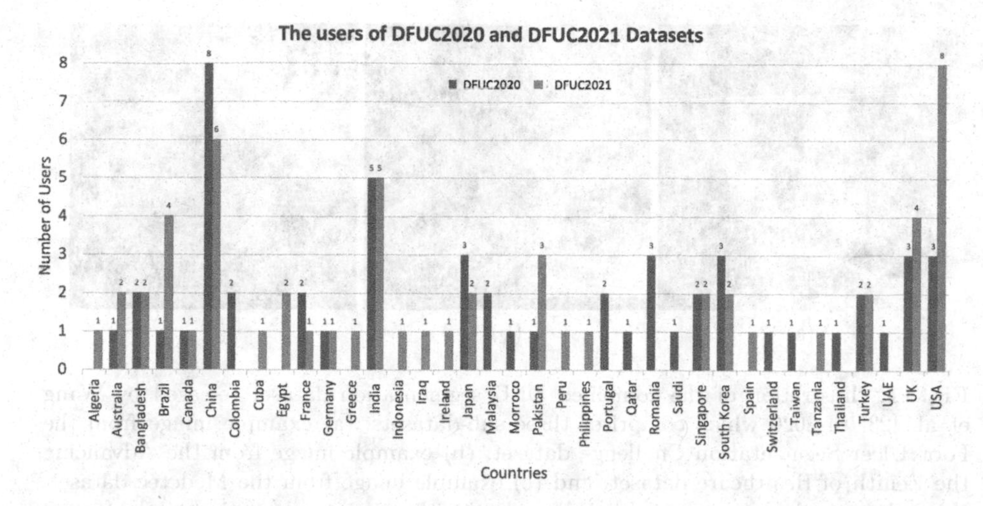

Fig. 10. Distribution of researchers using the DFUC2020 and DFUC2021 datasets and their country of origin. It is noted that DFUC2021 has reached wider research community, including researchers from Algeria, Cuba, Egypt, Greece, Indonesia, Iraq, Ireland, Peru, Philippines, Spain and Tanzania. Please note that we did not include New Zealand in this figure, as it was used by clinicians, rather than the challenge.

are used for testing. Images were captured in clinical settings at different angles, with many cases exhibiting background noise. The dataset contains images of the same ulcers at different angles at different time intervals. This dataset presents additional challenges for computer vision and deep learning algorithms as all images are padded with black pixels to maintain the aspect ratio and image size of 512×512 pixels, as illustrated in Fig. 11(a).

The AZH Wound Care Center Dataset was also introduced by Wang et al. in 2020 and contains 831 training images and 278 test images. All images in this dataset are 224×224 pixels and contain only DFU patches. The majority of each image in this dataset contains padding using black pixels, as shown in Fig. 11(b). A smaller dataset was also released (Medetec), as shown in Fig. 11(c). The release date of this dataset is unknown. This is a collection of multiple wound types, and contains 46 DFU images with no labels. The image resolution of this dataset ranges from 560×347 pixels to 224×444 pixels.

The segmentation dataset introduced by Wang et al. for the Foot UlcerSegmentation Challenge is bundled with 2 other datasets. First, is the Medetec wound dataset, resized to 224×224 with black padding along the bottom. The Medetec dataset has 160 wound images, 152 labels and 46 DFU examples. These images focus on the wounds, but still exhibit some background details. The second dataset is the AZH wound care center dataset, which has 1,109 images and 831 labels. Similar to Medetec, these images are 224×224, but padded at the bottom and sides with black borders. However, these images where pre-cropped to focus on the lesion only so do not show background on the rest of the foot.

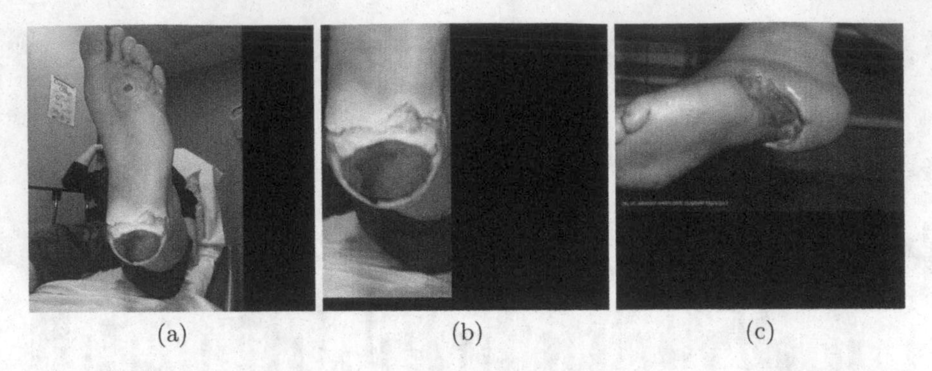

(a) (b) (c)

Fig. 11. Illustration of the combined DFU segmentation dataset released by Wang et al. [23] in 2020 which comprises three sub-datasets: (a) example image from the Foot Ulcer Segmentation Challenge dataset, (b) example image from the Advancing the Zenith of Healthcare dataset, and (c) example image from the Medetec dataset. Note that the images required padding due to the non-standard size of the source images.

4 DFU Challenges

Once a year, The Medical Image Computing and Computer Assisted Intervention (MICCAI) Society conduct medical image challenges[4] to support and lead to thoughtful research challenges. The registered MICCAI challenge is reviewed and evaluated by expert panels, criteria include the design, metrics and transparency toward higher quality challenges. Due to limited capacity and an increased number of proposals, some challenges were accepted as MICCAI endorsed events, which are online only challenges and are not associated with the conference.

The inaugural DFU challenge was initiated by Yap et al. [12] on the DFU detection task (DFUC2020). Lead by the Manchester Metropolitan University (UK) and Lancashire Teaching Hospitals (UK), together with other co-organisers including the University of Southern California (USA), University of Waikato (New Zealand), University of Manchester and Manchester Royal Infirmary (UK), Manipal College of Health Professions (India), Baylor College of Medicine (USA) and Waikato District Health Board (New Zealand). DFUC2020 was accepted as a MICCAI registered challenge, and was conducted in conjunction with MICCAI 2020. The DFUC2020 datasets [15] include 2,000 training images and 2,000 testing images. The summary of the challenge results were concluded by Yap et al. [31]. The organisers continue to support the research community with a live leaderboard on the Grand Challenge System[5]. To date, the best result on the live leaderboard reports an mAP of 0.73.

DFUC2021 [13] was accepted as a MICCAI registered challenge, and was conducted in conjunction with MICCAI 2021. The focus on DFUC2021 was on

[4] http://www.miccai.org/special-interest-groups/challenges/miccai-registered-challenges/.

[5] https://dfu2020.grand-challenge.org.

classification, where the DFU patches were classified into control/none, infection, ischaemia and both conditions [26]. The organisers continue to support the research community with a live leaderboard on the Grand Challenge System [6]. At the time of writing this paper, the best macro F1-score on the live leaderboard is 0.6307 [32].

In the same period, another research group based in the US organised an online-only Foot Ulcer Segmentation Challenge (FUSeg) [25], which was conducted as a MICCAI endorsed event. The best performance of FUSeg is a Dice score of 0.8880. Since the evaluation is not on a live leaderboard and the participants were required to send their codes (docker/container) to the organiser. It is currently unclear if the organisers are still accepting submissions.

5 Future Directions

In 2021, the DFUC2022 challenge proposal was accepted by MICCAI as a registered challenge [33]. The task concerns DFU segmentation[7]. Compared to the online-only segmentation challenge conducted in 2021, DFUC2022 will comprise of large-scale higher resolution datasets, as illustrated in Fig. 12. It will be conducted in conjunction with MICCAI 2022.

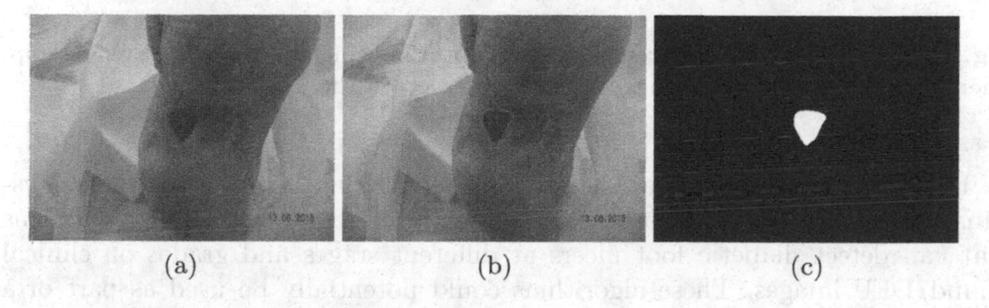

(a) (b) (c)

Fig. 12. Illustration of an example from DFUC2022 segmentation dataset, to be released in 2022, where: (a) sample image with higher resolution, (b) delineation by the expert, and (c) the binary mask for the ulcer's region.

To monitor the healing progress of DFUs in clinical settings, podiatrists/consultants take photographs of the foot using standard SLR cameras. Due to a lack of standardisation, the photographs taken are operator dependent, with variations in angle, distance and illumination. Figure 13 shows photographs of the same wound taken at different time points.

The efforts shown in previous research [7] attempted to standardise the data capturing process, enabling improved observation on images captured at different timelines. Such longitudinal datasets, as illustrated in Fig. 14, help computer vision techniques and human observers to better spot the subtle changes on feet.

[6] https://dfu-2021.grand-challenge.org.

[7] https://dfu-challenge.github.io/.

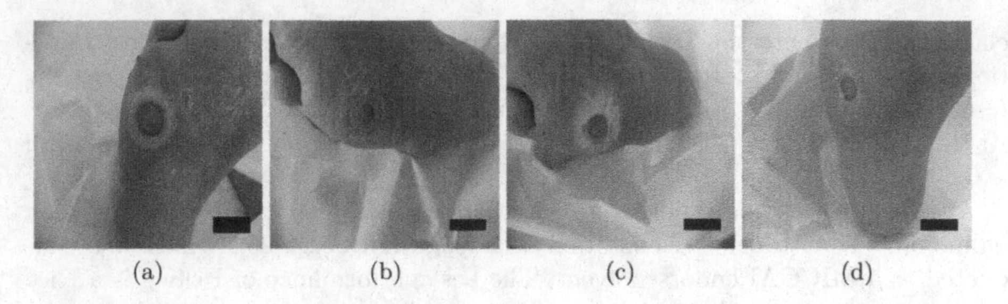

Fig. 13. Images acquired in a clinic at different intervals with difficulty in aligning the timeline photographs. This process is operator dependent and results in inconsistent illumination.

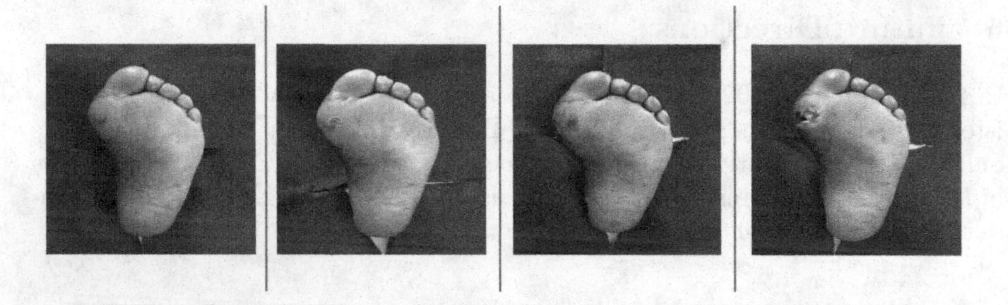

Fig. 14. Images taken at different intervals captured using the FootSnap mobile app, where subtle changes in the images can be observed on the forefoot.

In an effort to improve patient care and reduce the strain on healthcare systems, current research has been focused on the development of AI algorithms that can detect diabetic foot ulcers at different stages and grades on clinical wound/DFU images. These algorithms could potentially be used as part of a mobile application that patients could use themselves (or a carer/partner) to remotely monitor their foot condition and detect the appearance of DFU for timely clinical intervention. Also, clinicians and podiatrists can monitor the progress of DFU (detected by AI algorithms) through timeline images. This is an important step in deciding what therapeutic intervention is required depending upon the progression of DFU. Effective diagnosis of such wounds can lead to better treatments, which may lead to quicker healing, reduced amputation risk and a significant reduction in health care costs.

This paper discusses the challenges and opportunities in clinical DFU image datasets taken by different types of cameras and the role of AI algorithms in the detection and progress of DFU. Other than DFU images, researchers have used other imaging modalities such as infrared, Magnetic Resonance Imaging (MRI), and fluorescence imaging for the management of DFU. However, there are no such public datasets of other imaging modalities available for further research and development of AI solutions for multi-modality datasets.

In many research studies, thermal infrared imaging has been proven to be a useful technique in the clinical management of DFU. Several diabetic foot complications such as neuropathic ulcers, osteomyelitis and Charcot's foot have been identified at locations with increased temperature [34–37]. Increased plantar temperature is a strong indicator of pre-ulcer conditions and may present a week before an ulcer appears visually on the foot. Hence, regular monitoring of temperature (i.e. temperature difference ($>2.2\,^{\circ}$)) when comparing general foot temperature with suspected DFU sites can potentially help in early interventions.

Another imaging modality known as fluorescence imaging can detect the presence of clinically significant bacteria in diabetic foot ulcers by using a handheld device [38]. Potentially, fluorescence imaging can provide valuable information of DFU outcomes on whether a DFU is healing or not [39].

MRI is the modality of choice for imaging both Charcot foot and deep infection in the diabetic foot. In the early stages, MRI can demonstrate marrow oedema while plain films remain normal. the use of MRI is common with diabetic patients to rule out infection in the presence of an ulcer, to evaluate the severity of Charcot foot, or to distinguish between Charcot foot and infection [40–42].

In the current literature, there are no studies that combine AI interventions for the detection and management of DFUs in multi-modality imaging. Combining wound/ulcer images with different types of other imaging (such as MRI, Thermal Infrared, and fluorescence) can potentially help AI algorithms to provide a complete diagnosis and prognosis of DFUs and timely interventions in the treatment of diabetic foot ulcers to avoid amputations.

There are currently no publicly available datasets that combine the multi-modality imaging of diabetic foot patients. Great effort is needed for the collection of such datasets which combines different types of imaging such as thermal infrared (early detection of ulcers), clinical wound images (progression of ulcers), fluorescence (presence of clinically significant bacteria) and MRI (presence of infection and Charcot's foot). Similarly, most of the current state-of-the-art AI algorithms rely on supervised learning, hence annotation of the dataset is another important step required for the development of AI algorithms for diagnosis and management of the diabetic foot. DFU datasets are prone to the same issues that have affected other medical imaging datasets, such as image duplication, feature over-representation and sourcing of large numbers of images from a relatively small pool of subjects [43–45].

6 Conclusion

This paper provides an overview of the development of DFU datasets and notable advances in the field that have led to the current use of deep learning techniques. The aim is to guide researchers in this domain to understand the breadth and depth of the processes involved in DFU classification, detection and segmentation, and to promote good practice in research and data sharing.

Collection, labelling and curation of DFU datasets is a challenging process requiring significant input from clinical experts at all stages of development.

Comprehensive inter- and intra-rater analysis will prove to be key in refining the quality of datasets together with establishing new standards in this relatively new research domain. Researchers training deep learning models should pay particular attention to challenging examples within the datasets, as these will help to make networks more robust in real-world settings.

Acknowledgment. We gratefully acknowledge the support of NVIDIA Corporation who provided access to GPU resources for the DFUC2020 and DFUC2021 Challenges.

References

1. Armstrong, D.G., Lavery, L.A., Harkless, L.B.: Validation of a diabetic wound classification system: the contribution of depth, infection, and ischemia to risk of amputation. Diabetes Care **21**(5), 855–859 (1998)
2. Prompers, L., et al. Delivery of care to diabetic patients with foot ulcers in daily practice: results of the Eurodiale study, a prospective cohort study. Diabet. Med. **25**(6), 700–707 (2008)
3. Cavanagh, P., Attinger, C., Abbas, Z., Bal, A., Rojas, N., Zhang-Rong, X.: Cost of treating diabetic foot ulcers in five different countries. Diabet/Metab. Res. Rev. **28**(S1), 107–111 (2012)
4. Zimmet, P.Z., Magliano, D.J., Herman, W.R., Shaw, J.E.: Diabetes: a 21st century challenge. Lancet Diabet. Endocrinol. **2**(1), 56–64 (2014)
5. Vinicor, F.: The public health burden of diabetes and the reality of limits. Diabetes Care **21**(Supplement 3), C15–C18 (1998)
6. Chanussot-Deprez, C., Contreras-Ruiz, J.: Telemedicine in wound care: a review. Adv. Skin Wound Care **26**(2), 78–82 (2013)
7. Yap, M.H., et al.: A new mobile application for standardizing diabetic foot images. J. Diabetes Sci. Technol. **12**(1), 169–173 (2018)
8. Yap, M.H., et al.: Computer vision algorithms in the detection of diabetic foot ulceration a new paradigm for diabetic foot care? J. Diabetes Sci. Technol. **10**(2), 612–613 (2015)
9. Goyal, M., Yap, M.H., Reeves, N.D. Rajbhandari, S., Spragg, J.: Fully convolutional networks for diabetic foot ulcer segmentation. In: 2017 IEEE International Conference on Systems, Man, and Cybernetics (SMC), pp. 618–623, October 2017
10. Goyal, M., et al.: DFUNet: convolutional neural networks for diabetic foot ulcer classification. IEEE Trans. Emerg. Topics Comput. Intell. **4**(5), 728–739 (2018)
11. Goyal, M., Reeves, N.D., Rajbhandari, S., Ahmad, N., Wang, C., Yap, M.H.: Recognition of ischaemia and infection in diabetic foot ulcers: dataset and techniques. Comput. Biol. Med. **117**, 103616 (2020)
12. Yap, M.H.: Diabetic foot ulcers grand challenge 2020, March 2020
13. Yap, M.H.: Diabetic foot ulcers grand challenge 2021, March 2020
14. Goyal, M., Yap, M.H.: Region of interest detection in dermoscopic images for natural data-augmentation. arXiv preprint arXiv:1807.10711 (2018)
15. Cassidy, B.: The DFUC 2020 dataset: analysis towards diabetic foot ulcer detection. touchREV. Endocrinol. **17**, 5–11 (2021)
16. Wang, L., Pedersen, P.C., Strong, D.M., Tulu, B., Agu, E., Ignotz,R.: Smartphone-based wound assessment system for patients with diabetes. IEEE Trans. Med. Eng. **62**(2), 477–488 (2015)

17. Wang, L., et al.: An automatic assessment system of diabetic foot ulcers based on wound area determination, color segmentation, and healing score evaluation. J. Diabetes Sci. Technol. **10**, 08 (2015)
18. Wang, L., Pedersen, P.C., Agu, E., Strong, D.M., Tulu, B.: Area determination of diabetic foot ulcer images using a cascaded two-stage SVM-based classification. IEEE Trans. Biomed. Eng. **64**(9), 2098–2109 (2017)
19. Brown, R., loderer, B., Si Da Seng, L., Lazzarini, P., van Netten, J.: Myfootcare: a mobile self-tracking tool to promote self-care amongst people with diabetic foot ulcers. In: Proceedings of the 29th Australian Conference on Computer-Human Interaction, OZCHI 2017, pp. 462–466. Association for Computing Machinery, New York (2017)
20. Swerdlow, M., Shin, L., D'Huyvetter, K., Mack, W.J., Armstrong, D.G.: Initial clinical experience with a simple, home system for early detection and monitoring of diabetic foot ulcers: The foot selfie. J. Diabetes Sci. Technol. (2021)
21. Cassidy, B.: A cloud-based deep learning framework for remote detection of diabetic foot ulcers. arXiv preprint arXiv:2004.11853 (2021)
22. Reeves, N.D., Cassidy, B., Abbott, C.A., Yap, M.H.: Chapter 7 - novel technologies for detection and prevention of diabetic foot ulcers. In: Gefen, A. (ed.), The Science, Etiology and Mechanobiology of Diabetes and its Complications, pp. 107–122. Academic Press (2021)
23. Wang, C.: Fully automatic wound segmentation with deep convolutional neural networks. Sci. Rep. **10**(1), 1–9 (2020)
24. Steve Thomas. Medetec (2020). Accesed 08 Nov 2021
25. Wang, C., Rostami, B., Niezgoda, J., Gopalakrishnan, S., Yu, Z.: Foot ulcer segmentation challenge 2021, March 2021
26. Yap, M.H., Cassidy, B., Pappachan, J.M., O'Shea, C., Gillespie, D., Reeves, N.D.: Analysis towards classification of infection and ischaemia of diabetic foot ulcers. In: Proceedings of the IEEE EMBS International Conference on Biomedical and Health Informatics (BHI 2021), pp. 1–4 (2021)
27. Hewitt, B., Yap, M.H., M., Grant, H.: Manual whisker annotator (MWA): a modular open-source tool. J. Open Res. Softw. **4**(1) (2016)
28. Al-Garaawi, N., Ebsim, R., Alharan, A.F.H., Yap, M.H.: Diabetic foot ulcer classification using mapped binary patterns and convolutional neural networks. Comput. Biol. Med. **140**:105055 (2022)
29. Goyal, M., Reeves, N.D., Rajbhandari, S., Yap, M.H.: Robust methods for real-time diabetic foot ulcer detection and localization on mobile devices. IEEE J. Biomed. Health Inform. **23**(4), 1730–1741 (2019)
30. Dutta, A., Gupta, A., Zissermann, A.: VGG image annotator (VIA). https://github.com/ox-vgg/via, (2016). Version: 2.0.10., Accessed July 2020
31. Yap, M.H.: Deep learning in diabetic foot ulcers detection: a comprehensive evaluation. Comput. Biol. Med. **135**:104596 (2021)
32. Cassidy, B.: Diabetic foot ulcer grand challenge 2021: Evaluation and summary. arXiv preprint arXiv:2111.10376 (2021)
33. Yap, M.H.: Diabetic foot ulcers grand challenge 2022, March 2021
34. Harding, J.R., Wertheim, D.F., Williams, R.J., Melhuish, J.M., Banerjee, D., Harding. V.I.: Infrared imaging in diabetic foot ulceration. In: Proceedings of the 20th Annual International Conference of the IEEE Engineering in Medicine and Biology Society, vol. 20 Biomedical Engineering Towards the Year 2000 and Beyond (Cat. No. 98CH36286), vol. 2, pp. 916–918. IEEE (1998)

35. van Netten, J.J., van Baal, J.G., Liu, C., van Der Heijden, F., Bus, S.A.: Infrared thermal imaging for automated detection of diabetic foot complications. J. Diabetes Sci. Technol. **7**(5), 1122–1129 (2021)
36. Harding, J.R., Banerjee, D., Wertheim, D.F., Williams, R.J., Melhuish, J.M., Harding. K.G.: Infrared imaging in the long-term follow-up of osteomyelitis complicating diabetic foot ulceration. In: Proceedings of the First Joint BMES/EMBS Conference. 1999 IEEE Engineering in Medicine and Biology 21st Annual Conference and the 1999 Annual Fall Meeting of the Biomedical Engineering Society (Cat. No.), vol. 2, p. 1104. IEEE (1999)
37. Armstrong, D.G., Boulton, A.J.M., Bus, S.A.: Diabetic foot ulcers and their recurrence. New Engl. J. Med., **376**(24), 2367–2375 (2017)
38. Wu, Y.C.: Handheld fluorescence imaging device detects subclinical wound infection in an asymptomatic patient with chronic diabetic foot ulcer: a case report. Int. Wound J. **13**(4), 449–453 (2016)
39. John W Lindberg. Predicting clinical outcomes in a diabetic foot ulcer population using fluorescence imaging. Adv. Skin Wound Care **34**(11), 596–601 (2021)
40. Tan, P.L., Teh, J.: MRI of the diabetic foot: differentiation of infection from neuropathic change. Br. J. Radiol. **80**(959), 939–948 (2007)
41. Schwegler, B. et al.: Unsuspected osteomyelitis is frequent in persistent diabetic foot ulcer and better diagnosed by MRI than by 18F-FDG PET or 99mTc-MOAB. J. Int. Med. **263**(1), 99–106 (2008)
42. Forsythe, R.O., Hinchliffe, R.J.: Assessment of foot perfusion in patients with a diabetic foot ulcer. Diabetes/Metab. Res. Rev. **32**, 232–238 (2016)
43. Wen, D.: Characteristics of publicly available skin cancer image datasets: a systematic review. Lancet Digit. Health (2021)
44. Cassidy, B., Kendrick, C., Brodzicki, A., Jaworek-Korjakowska, J., Yap, M.H.: Usage, benchmarks and recommendations. Med. Image Anal. Anal. ISIC Image Datasets **75**(2021)
45. Daneshjou, R.: Checklist for evaluation of image-based artificial intelligence reports in dermatology: CLEAR Derm consensus guidelines from the international skin imaging collaboration artificial intelligence working group. JAMA Dermatol. **12** (2021
46. Abbott, C.A., et al.: Innovative intelligent insole system reduces diabetic foot ulcer recurrence at plantar sites: a prospective, randomised, proof-of-concept study. Lancet Digital Heal. **1**(6), e308–18 (2019)

DFUC 2021 Challenge Papers

Convolutional Nets Versus Vision Transformers for Diabetic Foot Ulcer Classification

Adrian Galdran[1,2](✉), Gustavo Carneiro[2], and Miguel A. González Ballester[1,3]

[1] BCN Medtech, Department of Information and Communication Technologies, Universitat Pompeu Fabra, Barcelona, Spain
[2] Australian Institute for Machine Learning, University of Adelaide, Adelaide, Australia
[3] ICREA, Barcelona, Spain

Abstract. This paper compares well-established Convolutional Neural Networks (CNNs) to recently introduced Vision Transformers for the task of Diabetic Foot Ulcer Classification, in the context of the DFUC 2021 Grand-Challenge, in which this work attained the first position. Comprehensive experiments demonstrate that modern CNNs are still capable of outperforming Transformers in a low-data regime, likely owing to their ability for better exploiting spatial correlations. In addition, we empirically demonstrate that the recent Sharpness-Aware Minimization (SAM) optimization algorithm improves considerably the generalization capability of both kinds of models. Our results demonstrate that for this task, the combination of CNNs and the SAM optimization process results in superior performance than any other of the considered approaches.

Keywords: Diabetic Foot Ulcer Classification · Vision Transformers · Convolutional Neural Networks · Sharpness-Aware Optimization

1 Introduction and Problem Statement

Diabetes affects 425 million people with a projection to reach 629 million affected people by 2045 [7]. A particularly concerning development of diabetes are foot ulcers: the risk of a diabetic patient developing a foot ulcer along their lifetime is estimated to be 19–34% [2].

Diabetic Foot Ulcers (DFU) presented with ischaemia and/or infections can lead to limb gangrene, amputation, or even patient death. Iscahemia is characterized by a deficient blood circulation, and it can be recognized by poor vascular reperfusion or dark gangrenous patterns. On the other hand, infection is associated to the presence of tissue inflammation or purulence [13]. Although continuous monitoring and timely diagnosis is crucial to avoid these serious complications, inappropriate care and lack of screening on diabetic population, as well as the lack of medical specialists in developing countries poses a barrier

M. H. Yap et al. (Eds.): DFUC 2021, LNCS 13183, pp. 21–29, 2022.
https://doi.org/10.1007/978-3-030-94907-5_2

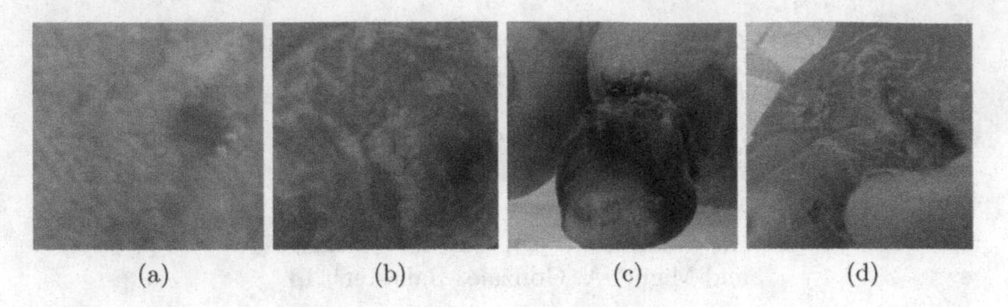

Fig. 1. Randomly sampled diabetic foot images from the DFUC 2021 dataset. (a) No infection and no ischaemia, (b) presence of infection, (c) presence of ischaemia, (d) presence of both conditions. Images provided by the organization.

towards early discovery and appropriate management of DFU. In this setting, computer-supported diagnosis by means of computer vision techniques and conventional smartphone cameras represents a promising and economically viable alternative, with some recent research focusing already on developing this kind of technology [5,13,26,27]. Therefore, previous research has addressed the need of automatic image understanding towards DFU analysis. Ranging from conventional feature engineering-based image classification [23] based on colour and texture to more advanced two-stage approaches involving lesion detection or segmentation followed by classification [1,4], and more recently deep learning approaches [8,18]. A detailed recent review on machine learning applied to DFU can be found in [21].

Within this context, the Diabetic Foot Ulcer Challenge (DFUC) 2021 was held in conjunction with the MICCAI 2021 conference. The task at hand is multi-class classification of foot images into four categories: no pathology, infection, ischaemia, or presence of both pathologies, as illustrated in Fig. 1. This paper describes our approach to solving the challenge, in which we obtained the first position. In addition, we make to contributions: 1) we compare the performance of two popular Convolutional Neural Networks (CNNs) [14,19] with that of two more recent Vision Transformers [10,20] in this problem, revealing that CNNs may still be the preferred choice for this kind of problems, and 2) we experiment with the Sharpness-Aware (SAM, [11]) extension of the popular Stochastic Gradient Descent optimization process, demonstrating that SAM yields superior performance in all the analyzed cases. More details on the DFUC 2021 organization, as well as live test leaderboard, can be found online at https://dfu-2021.grand-challenge.org/ .

In the next section, we first describe the two CNN and the two Transformer architectures we train for solving the task at hand. We then briefly recall the SAM optimization algorithm, and provide a description of the training process. Next, we present the data used for training and validating our analysis, and the performance obtained by each of the considered approaches, as well as the final test performance of our best model. We conclude the paper with some discussion and further research directions.

2 Methodology

We now describe the main methodological aspects of DFU classification in this paper: CNN and Vision Transformer architectures, model optimization process and training data.

2.1 Convolutional Neural Networks and Vision Transformers

In this work we experiment with two different kind of neural networks for computer vision, namely conventional Convolutional Neural Networks (CNNs) and more recent Vision Transformers. CNNs have become the de-facto approach to most computer vision tasks in the last decade, since their initial dominance of the 2012 ImageNet challenge [16]. Essential to the learning ability of CNNs from image data are several distinctive features, namely their built-in inductive bias (spatial translation invariance imposed through convolutions with shared learnable filters across the image), or a representation hierarchy arising from their layer-wise architecture design, by which larger patterns are learned as sequential combinations of smaller patterns. On the other hand, Transformers have emerged in the last few years as the most powerful architecture for Natural Language Processing applications, dominating virtually all public benchmarks[1] Recently, the Transformer architecture has been extended to other domains and data types, like graphs, speech, or vision [9,10,28]. After the initial exploration of Transformers for computer vision in [10], where a variety of image recognition tasks were successfully approached with this architecture, the amount of research dedicated to proposing Vision Transformers as an alternative to CNNs has increased exponentially. Generally speaking, Vision Transformers consider images as sequences of small patches akin to words or tokens, that can be supplied to standard Transformers. However, and in contrast with CNNs, Transformers have no notion of distance within an image, since patches are processed sequentially. Spatial relationships need to be modeled by means of positional encodings/embeddings and learned on extremely large datasets, in which case the increased flexibility of Transformers could favor them over CNNs. In this paper, we test Transformers' ability to generalize for a relatively small dataset of lesion images like those shown on Fig. 1. Specifically, the four architectures we will analyze for our problem are enumerated below.

- **Big Image Transfer (BiT)** [14]. We use the ResNeXt50 architecture from this work, which contains 25,55M parameters, uses Group Normalization instead of BatchNorm, and implements convolutional layers with Weight Standardization.
- **EfficientNet** [19], which is a family of architectures designed by neural architecture search, and carefully scaled so that they achieve state-of-the-art performance with a fraction of parameters. In this paper we use the EfficientNet B3 architecture, which contains 12,23M weights.

[1] Transformers are based on the notion of attention, the detailed description of which falls beyond the scope of this paper [22].

- **Vision Transformers (ViT)** [10], the first architecture to replace convolutions by attention, this is a pure transformer applied on sequences of image patches. We use the ViT-base configuration, containing 22,20M weights.
- **Data-efficient Image Transformers (DeIT)** [20], a refinement of ViT with improved pre-training strategies. We use the DeiT-small variant, which contains 22M learnable parameters, and again a patch size of 16×16.

2.2 Sharpness-Aware Optimization and Training Details

In this paper we are also interested in comparing standard Stochastic Gradient Descent (SGD, [3]) to a recently introduced approach for neural network optimization that tries to improve their generalization ability, namely Sharpness-Aware Optimization (SAM, [11,15]) . SAM simultaneously minimizes the value of a loss function and the loss sharpness, thereby seeking to find parameters lying in neighborhoods with a uniformly low loss. This way, the goal of SAM is to improve model generalization, and it has been shown to deliver State-of-the-Art performance for different applications, while also enforcing robustness against label noise. Further technical details about this optimization algorithm can be found in [11].

Other than the optimization algorithm choice, all the models trained in this work follow the same process, similarly to [12]: network weights are optimized so as to minimize the cross-entropy loss with each of the optimizers adopting an initial learning rate of $l = 0.01$ and a batch-size of 32. Note that these two values were selected by grid search on an initial train-validation split. The learning rate is decayed following a cosine law from its initial value to $l = 1e\text{-}8$ during 3 epochs, which defines a training cycle. This cycle is repeated 25 times, restarting the learning back at the beginning. During training images are augmented with standard techniques (random rotations, vertical/horizontal flipping, contrast/saturation/brightness changes). The F1-score is monitored on an independent validation set and the best performing model is kept for testing purposes. For testing, we generate four different versions of each image by horizontal/vertical flipping, predict on each of them, and average the results.

Note that in all cases, optimization starts from weights pre-trained on the ImageNet classification task[2] at a 224×224 resolution, which happens to coincide with the size of the image patches provided in the DFUC 2021 challenge, as explained in the next section.

2.3 Training Data

In DFUC 2021, participants were provided with a dataset composed of 15,683 DFU RGB image patches of a fixed 224×224 resolution, with 5,955 training and 5,734 testing unlabelled DFU patches for final ranking purposes. An additional set of 3,994 unlabeled DFU patches was released for training, although it was unused in this work. These patches were extracted from close-ups of diabetic feet

[2] All architectures and pretrained weights are taken from [24].

at a distance of approximately 30–40 cm perpendicularly to the plane of the ulcer. Natural data augmentation was included (repeated takes of the same lesion), and controlled for with automatic image similarity techniques while splitting the data into training and test subsets. More details on this dataset can be found in [25].

3 Experimental Results and Analysis

Below we analyze the performance of the four architectures described in Sect. 2.1 in the context of the DFUC 2021. The organizers provided participants with a labeled training set and offered an unlabeled validation set as well as a public leaderboard to which preliminary submissions could be submitted. After the validation period was over, the participants had to submit their predictions on a second hidden test set, and the final ranking was based on the latter. In all cases, the models described below were trained on a 4-fold split of the labeled training data, with predictions generated by average ensembling. Models were implemented in PyTorch [17] and trained on a workstation equipped with a NVIDIA GeForce RTX 3080 with 10 GB memory size.

3.1 Impact of SAM Optimization

In this section we compare the performance of the aforementioned four architectures when trained as specified above, employing both standard SGD and Sharpness-Aware loss minimization with SGD as base optimizer. Performance measures include F1-score (the metric that drives the DFU competition), as well as AUC, recall and precision on its macro versions. For this experiment, these quantities are measured on the intermediate validation set released by the organizers. Table 1 shows such results for all cases.

3.2 CNNs vs Transformers

We also study the performance of each of our four architectures in the final hidden test set. For this, Table 2 shows the metrics obtained by each model, whereas Table 3 describes the result of our final submission for the challenge classification, compared to the results obtained by the top-ranked contestants.

3.3 Discussion of the Results

The first conclusion we can draw from the above results is that the SAM optimization generally benefits the end performance of both CNNs and Transformers, with CNNs always appearing to be superior to Transformers in all cases. It should be noted that SAM requires two forward passes instead of one before weight update, which turns it into a relatively expensive alternative to SGD. However, results in Table 1 appear to indicate that the extra computational load is widely compensated, particularly when working with small datasets for which the cost of one training run is not substantial.

Table 1. Performance analysis of different combinations of architectures with SGD and SAM optimization procedures. Note that all metrics are macro-averaged across categories.

Model/Performance	F1-score	AUC	Recall	Precision
BiT-ResNeXt50 SGD	51.34	84.79	54.99	51.05
BiT-ResNeXt50 SAM	57.71	87.68	61.88	57.74
Perf. Diff.	**+6.37**	**+2.89**	**+6.89**	**+6.69**
EfficientNet B3 SGD	47.55	82.92	49.49	48.47
EfficientNet B3 SAM	57.65	84.81	60.74	57.13
Perf. Diff.	**+10.05**	**+1.89**	**+11.25**	**+8.66**
ViT SGD	48.84	85.35	53.13	51.07
ViT SAM	51.94	87.85	56.52	54.38
Perf. Diff.	**+3.10**	**+2.50**	**+3.39**	**+3.31**
DeiT SGD	53.87	84.92	58.01	53.01
DeiT SAM	53.98	85.97	58.12	54.03
Perf. Diff.	**+0.11**	**+1.05**	**+0.11**	**+1.02**

Table 2. Performance analysis in the final test set for the four different architectures considered in our comparison. Note that all metrics are macro-averaged across categories.

Model/Performance	F1-score	AUC	Recall	Precision
BiT-ResNeXt50	**61.53**	**88.49**	**65.59**	**60.53**
EfficientNet B3	59.71	87.01	61.79	59.34
ViT	58.48	87.64	62.27	58.91
DeiT	57.29	87.98	61.35	57.01

Next, we can see in Table 2 that, as expected, the relative ranking of our four models in the validation set is preserved when evaluating them in the final test set. However, there is a noticeable increase in the performance of Vision Transformers, which attain results close to those of the EfficientNet B3 architecture. It is also worth stressing that, just as in the validation experiments on Table 1, the BiT-ResneXt50 model still achieves the highest performance, surpassing by a good margin the second best model, which is EfficientNet B3.

Finally, we see from the final ranking of the DFUC 2021 challenge[3] in Table 3 that our approach is validated when compared to other contestants' results. Our final highest score was achieved by a linear combination of the predictions extracted from BiT-ResneXt50 and EfficientNet B3, but it is worth noting that

[3] Ongoing (live) leaderboard can be accessed at https://dfu-2021.grand-challenge.org/evaluation/live-testing-leaderboard/leaderboard/.

Table 3. Performance analysis in the final test set for the five top competitors sorted by ranking in terms of F1-score. Note that our submission in this table corresponds to an ensemble of the predictions from the BiT-ResNeXt50 and EfficientNet B3 models in Table 2. Note that all metrics are macro-averaged across categories.

Participant/Performance	F1-score	AUC	Recall	Precision
Agaldran (this work)	**62.16**	**88.55**	**65.22**	61.40
Louise.Bloch	60.77	86.16	62.46	**62.07**
S. Ahmed	59.59	86.44	59.79	59.84
Abdul	56.91	84.88	61.04	58.14
Orhunguley	56.10	87.02	57.59	59.17

BiT-ResneXt50 alone would still be the ranked the best method, and EfficientNet B3 would have been ranked second.

4 Conclusion

This paper describes the winning solution to the DFUC 2021 grand-challenge on Diabetic Foot Ulcer Classification, held in conjunction with MICCAI 2021, which is based on an ensemble of CNNs (BiT-Resnext50 and EfficientNet B3) trained on different data folds. In addition, for weight optimization we employ Sharpness-Aware Minimization, which provides a noticeable improvement in performance for all considered models. Note that, in the literature, SAM has been benchmarked mostly with CNNs, which could introduce a bias of this optimization algorithm towards that option in this paper. However, it has been recently shown that SAM has equally good impact in performance for Vision Transformers [6]. Finally, we also include a comparison of the performance of CNNs and Vision Transformers for this task, which reveals that despite the recent popularity of Transformers for Computer Vision tasks, CNNs may still be the preferred choice in few-data scenarios.

Acknowledgments. Adrian Galdran was funded by a Marie Skłodowska-Curie Global Fellowship (No. 892297). Gustavo Carneiro was partially supported by Australian Research Council grants (DP180103232 and FT190100525).

References

1. Amin, J., Sharif, M., Anjum, M.A., Khan, H.U., Malik, M.S.A., Kadry, S.: An integrated design for classification and localization of diabetic foot ulcer based on CNN and YOLOv2-DFU models. IEEE Access **8**, 228586–228597 (2020). https://doi.org/10.1109/ACCESS.2020.3045732
2. Armstrong, D.G., Boulton, A.J.M., Bus, S.A.: Diabetic foot ulcers and their recurrence. New Engl. J. Med. **376**(24), 2367–2375 (2017). https://doi.org/10.1056/NEJMra1615439

3. Bottou, L., Bousquet, O.: The tradeoffs of large scale learning. In: Platt, J., Koller, D., Singer, Y., Roweis, S. (eds.) Advances in Neural Information Processing Systems. vol. 20. Curran Associates, Inc. (2008). https://proceedings.neurips.cc/paper/2007/file/0d3180d672e08b4c5312dcdafdf6ef36-Paper.pdf
4. Brüngel, R., Friedrich, C.M.: DETR and YOLOv5: exploring performance and self-training for diabetic foot ulcer detection. In: 2021 IEEE 34th International Symposium on Computer-Based Medical Systems (CBMS), pp. 148–153, June 2021. https://doi.org/10.1109/CBMS52027.2021.00063, ISSN:2372-9198
5. Cassidy, B., et al.: The DFUC 2020 dataset: analysis towards diabetic foot ulcer detection. Touch Rev. Endocrinol. **17**, 5–11 (2021). https://doi.org/10.17925/EE.2021.17.1.5
6. Chen, X., Hsieh, C.J., Gong, B.: When vision transformers outperform ResNets without pre-training or strong data augmentations. arXiv:2106.01548 [cs] (Oct 2021), http://arxiv.org/abs/2106.01548, arXiv: 2106.01548
7. Cho, N.H., et al.: IDF diabetes atlas: global estimates of diabetes prevalence for 2017 and projections for 2045. Diabet. Res. Clin. Pract. **138**, 271–281 (2018). https://doi.org/10.1016/j.diabres.2018.02.023
8. Das, S.K., Roy, P., Mishra, A.K.: Recognition of ischaemia and infection in diabetic foot ulcer: a deep convolutional neural network based approach. Int. J. Imag. Syst. Technol. https://doi.org/10.1002/ima.22598. _eprint: https://onlinelibrary.wiley.com/doi/pdf/10.1002/ima.22598
9. Dong, L., Xu, S., Xu, B.: Speech-transformer: a no-recurrence sequence-to-sequence model for speech recognition. In: 2018 IEEE International Conference on Acoustics, Speech and Signal Processing (ICASSP), pp. 5884–5888, April 2018. https://doi.org/10.1109/ICASSP.2018.8462506, ISSN:2379-190X
10. Dosovitskiy, A., et al.: An image is worth 16 × 16 words: transformers for image recognition at scale. In: International Conference on Learning Representations (2021)
11. Foret, P., Kleiner, A., Mobahi, H., Neyshabur, B.: Sharpness-aware minimization for efficiently improving generalization. In: International Conference on Learning Representations (2021). https://openreview.net/forum?id=6Tm1mposlrM
12. Galdran, A., Anjos, A., Dolz, J., Chakor, H., Lombaert, H., Ayed, I.B.: The Little W-Net that could: state-of-the-art retinal vessel segmentation with minimalistic models, September 2020. http://arxiv.org/abs/2009.01907
13. Goyal, M., Reeves, N.D., Rajbhandari, S., Ahmad, N., Wang, C., Yap, M.H.: Recognition of ischaemia and infection in diabetic foot ulcers: Dataset and techniques. Comput. Biol. Med. **117**, 103616 (2020)
14. Kolesnikov, A., et al.: Big transfer (BiT): general visual representation learning. In: ECCV (2020)
15. Korpelevich, G.M.: The extragradient method for finding saddle points and other problems. Ekonomika i Matematicheskie Metody **12**, 747–756 (1976)
16. Krizhevsky, A., Sutskever, I., Hinton, G.E.: ImageNet classification with deep convolutional neural networks. In: Pereira, F., Burges, C.J.C., Bottou, L., Weinberger, K.Q. (eds.) Advances in Neural Information Processing Systems. vol. 25. Curran Associates, Inc. (2012), https://proceedings.neurips.cc/paper/2012/file/c399862d3b9d6b76c8436e924a68c45b-Paper.pdf
17. Paszke, A., et al.: An imperative style, high-performance deep learning library. In: Wallach, H., Larochelle, H., Beygelzimer, A., Alché-Buc, F.D., Fox, E., Garnett, R. (eds.) Advances in Neural Information Processing Systems 32, pp. 8024–8035. Curran Associates, Inc. (2019), http://papers.neurips.cc/paper/9015-pytorch-an-imperative-style-high-performance-deep-learning-library.pdf

18. Rostami, B., Anisuzzaman, D.M., Wang, C., Gopalakrishnan, S., Niezgoda, J., Yu, Z.: Multiclass wound image classification using an ensemble deep CNN-based classifier. Comput. Biol. Med. **134** (2021). https://doi.org/10.1016/j.compbiomed.2021.104536
19. Tan, M., Le, Q.: EfficientNet: rethinking model scaling for convolutional neural networks. In: International Conference on Machine Learning, pp. 6105–6114. PMLR, May 2019
20. Touvron, H., Cord, M., Douze, M., Massa, F., Sablayrolles, A., Jegou, H.: Training data-efficient image transformers & distillation through attention. In: Proceedings of the 38th International Conference on Machine Learning, July 2021
21. Tulloch, J., Zamani, R., Akrami, M.: Machine learning in the prevention, diagnosis and management of diabetic foot ulcers: a systematic review. IEEE Access **8**, 198977–199000 (2020). https://doi.org/10.1109/ACCESS.2020.3035327
22. Vaswani, A., et al.: Attention is all you need. In: Guyon, I., et al. (eds.) Advances in Neural Information Processing Systems. vol. 30. Curran Associates, Inc. (2017). https://proceedings.neurips.cc/paper/2017/file/3f5ee243547dee91fbd053c1c4a845aa-Paper.pdf
23. Veredas, F., Mesa, H., Morente, L.: Binary tissue classification on wound images with neural networks and Bayesian classifiers. IEEE Trans. Med. Imag. **29**(2), 410–427 (2010). https://doi.org/10.1109/TMI.2009.2033595
24. Wightman, R.: PyTorch Image Models (2019)
25. Yap, M.H., Cassidy, B., Pappachan, J.M., O'Shea, C., Gillespie, D., Reeves, N.: Analysis towards classification of infection and ischaemia of diabetic foot ulcers, June 2021, http://arxiv.org/abs/2104.03068
26. Yap, M.H., et al.: A new mobile application for standardizing diabetic foot images. J. Diabet. Sci. Technol. **12**(1), 169–173 (2018). https://doi.org/10.1177/1932296817713761
27. Yap, M.H.: Computer vision algorithms in the detection of diabetic foot ulceration. J. Diabet. Sci. Technol. **10**(2), 612–613 (2015)
28. Yun, S., Jeong, M., Kim, R., Kang, J., Kim, H.J.: Graph transformer networks. In: Wallach, H., Larochelle, H., Beygelzimer, A., Alché-Buc, F.d., Fox, E., Garnett, R. (eds.) Advances in Neural Information Processing Systems, vol. 32. Curran Associates, Inc. (2019). https://proceedings.neurips.cc/paper/2019/file/9d63484abb477c97640154d40595a3bb-Paper.pdf

Boosting EfficientNets Ensemble Performance via Pseudo-Labels and Synthetic Images by pix2pixHD for Infection and Ischaemia Classification in Diabetic Foot Ulcers

Louise Bloch[1,2](✉), Raphael Brüngel[1,2], and Christoph M. Friedrich[1,2]

[1] Department of Computer Science, University of Applied Sciences and Arts Dortmund (FH Dortmund), Emil-Figge-Str. 42, 44227 Dortmund, Germany
{louise.bloch,raphael.bruengel,christoph.friedrich}@fh-dortmund.de
[2] Institute for Medical Informatics, Biometry and Epidemiology (IMIBE), University Hospital Essen, Hufelandstr. 55, 45122 Essen, Germany

Abstract. Diabetic foot ulcers are a common manifestation of lesions on the diabetic foot, a syndrome acquired as a long-term complication of diabetes mellitus. Accompanying neuropathy and vascular damage promote acquisition of pressure injuries and tissue death due to ischaemia. Affected areas are prone to infections, hindering the healing progress. The research at hand investigates an approach on classification of infection and ischaemia, conducted as part of the Diabetic Foot Ulcer Challenge (DFUC) 2021. Different models of the EfficientNet family are utilized in ensembles. An extension strategy for the training data is applied, involving pseudo-labeling for unlabeled images, and extensive generation of synthetic images via pix2pixHD to cope with severe class imbalances. The resulting extended training dataset features 8.68 times the size of the baseline and shows a real to synthetic image ratio of 1:3. Performances of models and ensembles trained on the baseline and extended training dataset are compared. Synthetic images featured a broad qualitative variety. Results show that models trained on the extended training dataset as well as their ensemble benefit from the large extension. F1-Scores for rare classes receive outstanding boosts, while those for common classes are either not harmed or boosted moderately. A critical discussion concretizes benefits and identifies limitations, suggesting improvements. The work concludes that classification performance of individual models as well as that of ensembles can be boosted utilizing synthetic images. Especially performance for rare classes benefits notably.

Keywords: Diabetic foot ulcers · Classification ensemble · Pseudo-labeling · Generative adversarial networks · EfficientNets · pix2pixHD

L. Bloch and R. Brüngel—These authors contributed equally to this work.

M. H. Yap et al. (Eds.): DFUC 2021, LNCS 13183, pp. 30–49, 2022.
https://doi.org/10.1007/978-3-030-94907-5_3

1 Introduction

In 2019 there was an estimated amount of 463 million diabetes mellitus cases (9.3% of the world's population) [24]. This number is expected to rise up to 578 million cases (10.2%) until 2030 [24]. Associated with the disease is the diabetic foot syndrome, a long-term complication that can manifest with neuropathy and ischaemia. Without proper monitoring and care, diabetic foot ulcers (DFUs) may arise from these, which have an estimated global prevalence of 6.3% in diabetics [40]. Impaired wound healing [9] and common complications such as infections [26] facilitate chronification, hence regular and attentive screening and documentation are necessitated. Deficient care can prolong treatment, cause aggravation, and ultimately make amputations necessary. Beside a resulting harsh impact on the quality of life, amputation wounds are again prone to complications.

To support overburdened caregivers and facilitate best practices, machine learning-based applications are a key technology. These enable automation of time-consuming tasks and provision of decision support at the point-of-care. This includes the early and certain recognition of adverse shifts in the wound healing progress such as infection and ischaemia. The DFU Challenge (DFUC) is a series of academic challenges that address tasks related to DFU care to enable a broad comparison of detection [35], classification [36], and segmentation [37] methods as well as to evaluate the state of the art [34] for potential applications.

The work at hand presents a contribution to the DFUC 2021 [36] on classification of infection and ischaemia in DFU images. It uses an EfficientNets [27] ensemble that achieved the 2nd place. Its models were trained on an extended and class-balanced dataset. This was established by via pseudo-labeling of unlabeled images and, as a novelty in DFU classification, via class-individual generation of synthetic images using pix2pixHD [31]. Related work on DFU classification was conducted by [1,5,12] to discriminate healthy and abnormal skin. Recent and strongly related work on classification of infection and ischaemia in DFU was addressed by [6,12]. Benchmark results for the DFUC 2021 were presented in [33]. Generation of synthetic wound images was priorly addressed by [25,39], yet not specifically for DFU.

The manuscript consent is organized as follows: In Sect. 2 descriptions on used data, methods, and the experiment environment are covered. The approach followed as well as the used experiment setup are elaborated in Sect. 3. Results achieved during the challenge are presented in Sect. 4 featuring visualizations for explainability, discriminating those without and with the use of pseudo-labels and synthetic images. Section 5 provides a critical discussion on the approach, results, and limitations. Eventually, Sect. 6 summarizes results and draws conclusions on the potential of the presented approach.

2 Data and Methods

In the following, the DFUC 2021 challenge dataset with its modalities is described. Further, EfficientNets and pix2pixHD as used methods as well as the environment experiments were performed in are elaborated.

2.1 Diabetic Foot Ulcer Challenge 2021 Dataset

The DFUC 2021 [36] dataset [33] focuses on identification and analysis of infection and ischaemia DFU images. It comprises four classes, showing neither infection nor ischaemia (`none`), either infection (`infection`) or ischaemia (`ischaemia`), or both combined (`both`). Data was collected from Lancashire Teaching Hospitals[1] in a non-laboratory environment. Hence, images comprise flaws such as blurring, poor lighting, and reflection artifacts. Experts extracted patches [33] with a resolution of 224×224 px containing DFU regions. The resulting dataset was split into a training and a test part, images of both partitions were augmented to generate additional data, excluding too similar images [33]. The overall process resulted in 15,683 images: 5,955 (37.97%) labeled training images, 3,994 (25.47%) unlabeled training images, and 5,734 (36.56%) test images. A validation dataset of 500 (8.72%) images was extracted from the test part. The labeled training part comprises 2,555 (42.91%) `infection` images, 227 (3.81%) `ischaemia` images, 621 (10.43%) `both` images, and 2,552 (42.85%) `none` images [33]. Figure 1 shows examples, provided by the maintainers.

Beside the low resolution of patches and the overlapping class `both`, the dataset features further obstacles. The risk of information leakage is present due to an unclear generation of original training and test sets which might not be split on the subject level. In addition, the choice of augmented image inclusion can be questioned as whether augmentations should rather be dedicated to challenge contestants, as model selection strategies are impacted by these.

2.2 Classification via EfficientNets

The EfficientNet[2] [27] base model is a classification network developed using a CNN architecture search. The search aims to optimize classification models for performance (measured in accuracy) and training time (measured in Floating Point Operations Per Second (FLOPS)) in parallel. To increase image resolution, model depth, and model width, this base model is gradually scaled up using a uniform balance. All models of the EfficientNet family (EfficientNet-B0 up to EfficientNet-B7) achieved state-of-the-art performances on the ImageNet [7] classification task using smaller and faster model architectures [27].

2.3 Image Synthesis via pix2pixHD

The pix2pixHD[3] [31] framework enables photo-realistic high-resolution image synthesis and image-to-image translation for images up to 2048×1024 px. It represents a refined version of pix2pix [15], based on a conditional [21] Generative Adversarial Network (GAN) [11] architecture, combining a novel and more

[1] Lancashire Teaching Hospitals: https://www.lancsteachinghospitals.nhs.uk/, access 2021-09-22.

[2] EfficientNet: https://github.com/mingxingtan/efficientnet, access 2021-10-03.

[3] pix2pixHD: https://github.com/NVIDIA/pix2pixHD, access 2021-09-12.

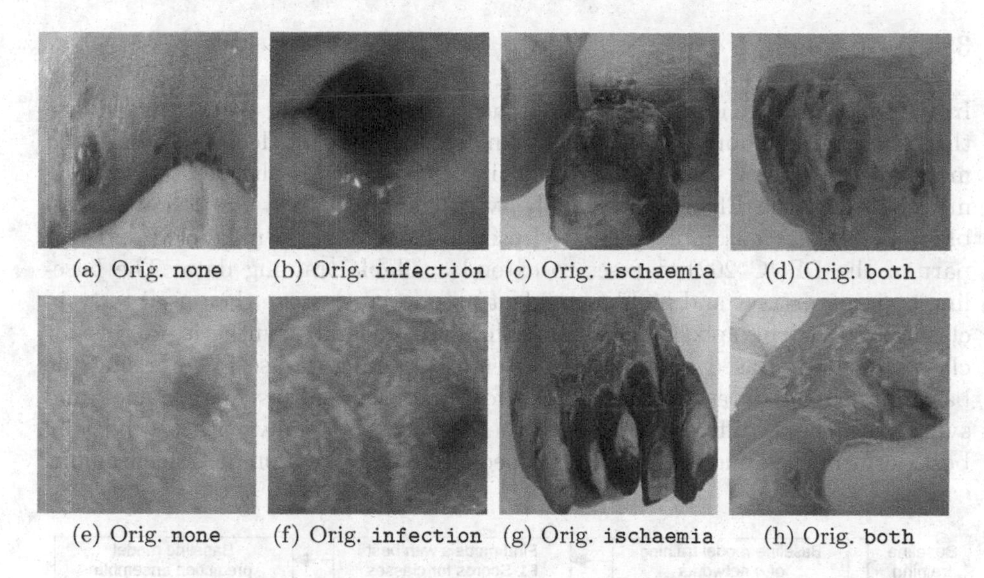

(a) Orig. none (b) Orig. infection (c) Orig. ischaemia (d) Orig. both

(e) Orig. none (f) Orig. infection (g) Orig. ischaemia (h) Orig. both

Fig. 1. Examples from the DFUC 2021 dataset for all classes.

robust adversarial learning objective with a multi-scale generator/discriminator [31]. Hereby, it addresses the problem of lacking details and realistic textures for high resolutions [15, 31]. pix2pixHD further features interactive semantic manipulation for objects on an instance level as well as generation of different synthetic images for a single input [31]. Beside the use of semantic label masks, it also allows training and generation via edge masks in a zero-class mode.

2.4 Experimental Environment

Experiments were conducted on NVIDIA® V100[4] tensor core Graphical Processing Units (GPUs) with 16 GB memory. These were part of an NVIDIA® DGX-1[5], a supercomputer specialized for deep learning. The operating system was Ubuntu Linux[6] in version `20.04.2 LTS (Focal Fossa)`, the driver version was `450.119.04`, and the used Compute Unified Device Architecture (CUDA) version was `10.1`. The execution environment was an NVIDIA®-optimized[7] Docker[8] [19] container engine, running a Deepo[9] image for a quick setup. Unless stated otherwise, experiments were conducted on a single GPU.

[4] V100: https://www.nvidia.com/en-us/data-center/v100/, access 2021-09-13.
[5] DGX-1: https://www.nvidia.com/en-us/data-center/dgx-1/, access 2021-09-13.
[6] Ubuntu Linux: https://ubuntu.com/, access 2021-07-10.
[7] NVIDIA®-Docker: https://github.com/NVIDIA/nvidia-docker, access 2021-07-10.
[8] Docker: https://www.docker.com/, access 2021-07-10.
[9] Deepo: https://github.com/ufoym/deepo, access 2021-09-22.

3 Approach

In the following, the implemented approach visualized in Fig. 2 and divided into three phases is elaborated. In the baseline phase, different deep learning-based models were trained on the baseline training dataset and the best performing models, all of the EfficientNet family, were combined to a prediction ensemble. The average ensemble generated pseudo-labels for the unlabeled and test part of the DFUC 2021 dataset to extend available training data. The baseline training dataset and highly confident pseudo-labels were then used to train class-individual pix2pixHD models, utilized to generate synthetic images for class-balancing. Based on this final extended training dataset, comprising the baseline training dataset, pseudo-labels for unlabeled and test part images, and synthetic images, different models of the EfficientNet family with the initially best performing configuration were trained and merged to a prediction ensemble.

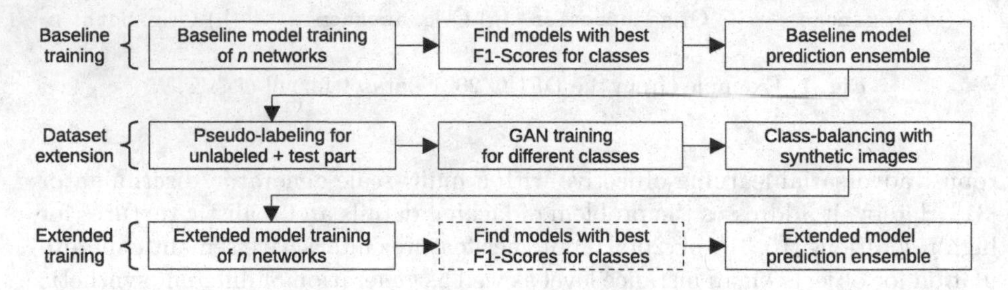

Fig. 2. Implemented three-phase workflow: Training with baseline data, extension of baseline data, and training with extended data. In the third phase, no F1-Score evaluation (dashed box) was possible due to expiration of the validation phase.

3.1 Baseline Models and Prediction Ensemble

During the validation stage of the challenge, explorative experiments were executed to investigate the performances of different deep learning-based models, including EfficientNets [27], EfficientNet-v2 [28], Vision Transformers [8] and ResNet 101 [13]. Those models were loaded using the Python package PyTorch image models (`timm`) [32] and trained using the Python package PyTorch [22] on the original training dataset. Further experiments were executed using different learning rates, numbers of epochs, optimizers, oversampling strategies, and a step based learning-rate scheduler. All models were pre-trained using the ImageNet-1k or ImageNet-21k dataset. For some models, a warm-up phase was implemented to first train the added classification layers. Cross-entropy loss was used for all experiments. For each model, the highest mini-batch size was determined. Mixed precision [20] was implemented to decrease memory requirements during the training process and consequently increase mini-batch size. The input images were 224×224 px. Two augmentation pipelines – one baseline pipeline and one extended pipeline – were implemented using the Albumentations [2] package. The baseline augmentation pipeline consists of basic augmentations: resizing, random cropping, vertical and horizontal flipping, geometrical shifting, scaling and rotation, an RGB shift, random brightness contrast and image normalization. The extended augmentation pipeline included resizing, random cropping, vertical and horizontal flipping, geometrical shifting, scaling and rotation, random brightness contrast, blurring and median blurring, downscaling, elastic transforms, optical distortions, grid distortions, and image normalization.

For all models, a 5-fold cross-validation (CV) was implemented. However, since the baseline training dataset contains augmented images [33], training and validation sets of CV-splits were not independent, leading to overestimated model performances. Those baseline models that reached the best class F1-Scores during the validation stage were combined to an average ensemble. Baseline model parameters are summarized in Table 1, results during the validation stage are summarized in Table 2. To improve model performances and generalizability [17], averaging was implemented without weights. The average ensemble was used to generate pseudo-labels for unlabeled images of training and test parts.

Table 1. Used hyperparameters to train the baseline models. All models used a dropout ratio of 0.3, were pre-trained for the ImageNet-1k dataset, and used an image size of 224×224 px.

Parameter	B_1	B_2	B_3	B_4
EfficientNet model architecture	B1	B0	B2	B1
Epochs warm-up	0	0	0	3
Learning rate warm-up	No	No	No	10^{-2}
Epochs training	30	100	30	47
Learning rate training	10^{-4}	10^{-4}	10^{-4}	10^{-4}
Batch size	225	300	225	225
Oversampling	No	Yes	No	No
Augmentations	Baseline	Extended	Baseline	Baseline
Optimizer	Adam	Adam	Adam	RMSprop
Learning rate scheduler	No	No	Step	Step
Step size	No	No	10	10
Gamma	No	No	0.1	0.1

3.2 Pseudo Labeling and Synthetic Image Generation

In the second phase, the baseline training dataset was extended in two steps: (i) Creation of pseudo-labels for not yet labeled images for initial extension, in particular for the underrepresented classes `ischaemia` and `both`, and (ii) generation of synthetic images to extend training data as well as to cope with class imbalances. Details of the resulting class distribution are listed in Table 3 and further elaborated in the following.

Table 2. Official classification results of the baseline models for the validation part of the dataset: Macro, weighted average (WA), and class F1-Scores (F1) as well as the Accuracy. Best results are highlighted.

Model	none F1 %	infection F1 %	ischaemia F1 %	both F1 %	Acc. %	WA F1 %	macro F1 %
B_1	71.02	58.64	35.90	**56.18**	63.60	**63.04**	**55.43**
B_2	68.57	57.29	**39.13**	52.50	61.60	61.15	54.37
B_3	**72.41**	52.00	38.89	45.65	61.60	59.79	52.24
B_4	71.05	**59.85**	37.04	49.41	**63.80**	**63.04**	54.34

For pseudo-labeling the model ensemble created in workflow phase 1 was used to infer predictions for the unlabeled and test part of the dataset. In sum, both dataset parts comprised 9,728 images (3,994 unlabeled, 5,734 test). To only use quite confident predictions, a confidence threshold of 70% for a single class was set as condition to ascribe an image to it. This was done to exclude rather uncertain predictions that would have been more likely to represent false-positive cases, having a negative impact on the classification performance of models trained on the extended dataset. A total of 6,961 predictions fulfilled the set requirement and were considered as pseudo-labeled training data extension. This way, the amount of images of the `ischaemia` class could be increased by 189 (+83.26%), and that of the `both` class by 321 (+51.69%). The amount of images for the `none` and `infection` classes could be increased as well by 4,348 (+170.38%) and 2,103 (+82.31%). Yet, their extension was less crucial for the second extension step due to an already sufficient amount of images. After the first step of pseudo-labeling, the `none` class comprised 6,900 images, `infection` 4,658 images, `ischaemia` 416 images, and `both` 942 images.

For synthetic image generation, individual pix2pixHD models for each class had to be created. As no area masks with regions of interest were available for images, edge masks were created for images of the extended dataset using the Canny edge detection algorithm [3] implemented in ImageMagick[10] version `6.9.10-23 Q16 x86_64 20190101`. The default parameterization was used, setting the radius to 0, the standard deviation to 1, and the percent level range to $[10, 30]$. Resulting edge masks enabled training in a zero-class mode, considering the whole image content with the aid of a respective sketch as a support structure. Individual pix2pixHD models were then trained on class-specific splits of the training dataset extended with pseudo-labeled images from the first step. Used parameters and settings are listed in Table 4. The chosen batch size was the maximum possible amount of images, limited by the GPU RAM, yet increased by using mixed precision. The default learning rate of $2 \cdot 10^{-4}$ was raised to $3 \cdot 10^{-4}$, as instabilities[11], occurring during early stages of training, were less likely to persist. The amount of epochs with the initial and decaying learning rate was chosen manually by observing intermediate results during

[10] ImageMagick: https://github.com/ImageMagick/ImageMagick, access 2021-09-22.
[11] pix2pixHD artifacts: https://github.com/NVIDIA/pix2pixHD/issues/46, access 2021-09-11.

training when synthetic images were decided to be sufficiently detailed and convincing. Trained models were then used to generate synthetic images. To cope with the considerable class imbalance, for each class synthetic images were created using the edge masks of all other classes. I.e., the 6,900 given `none` images were extended generating further 6,016 synthetic images via the `none` model, using the 4,658 `infection`, 416 `ischaemia`, and 942 `both` class edge masks and vice versa. Figure 3 illustrates the translation of a single edge mask of the `none` class (Fig. 3a) to three synthetic images of the `infection`, `ischaemia`, and `both` classes (Fig. 3b, Fig. 3c, and Fig. 3d). Hence, after the second step of synthetic image extension each class comprised 12,916 labeled images, summing up to 51,664 training images including baseline, pseudo-labeled and synthetic images.

Table 3. Proportions of the extended training dataset after two extensions.

Class	Baseline training data	Pseudo-label extension	Syn. image extension	Σ
`none`	2,552 (4.94%)	4,348 (8.42%)	6,016 (11.64 %)	12,916 (25.00%)
`infection`	2,555 (4.95%)	2,103 (4.07%)	8,258 (15.98%)	12,916 (25.00%)
`ischaemia`	227 (0.44%)	189 (0.37%)	12,500 (24.19%)	12,916 (25.00%)
`both`	621 (0.12%)	321 (0.62%)	11,974 (23.18%)	12,916 (25.00%)
Σ	5,955 (11.53%)	6,961 (13.47%)	38,748 (75.00%)	51,664 (100.00%)

Table 4. pix2pixHD parameters used for individual class model training.

Parameter/Setting	`none`	`infection`	`ischaemia`	`both`
Number of classes	0	0	0	0
Mixed precision	Yes	Yes	Yes	Yes
Batch size	48	48	48	48
Learning rate	$3 \cdot 10^{-4}$	$3 \cdot 10^{-4}$	$3 \cdot 10^{-4}$	$3 \cdot 10^{-4}$
Epochs with initial learning rate	50	50	200	200
Epochs with decaying learning rate	100	100	400	400
Load/fine size	224 px	224 px	224 px	224 px
Resize/crop	No	No	No	No
Instance maps	No	No	No	No

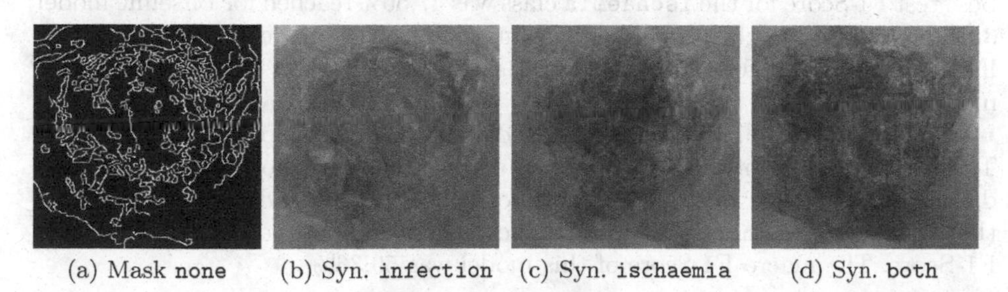

(a) Mask `none` (b) Syn. `infection` (c) Syn. `ischaemia` (d) Syn. `both`

Fig. 3. Examples for synthetic images generated via a mask of the `none` class for the `infection`, `ischaemia`, and `both` classes.

3.3 Extended Models and Prediction Ensemble

Based on the synthetic images generated in the second phase of the workflow, three models were trained with the deep learning-based classification pipeline of phase 1. To increase the mini-batch size and decrease the training time, the pipeline was trained using four GPU cores. Due to the expiration of the challenge's validation phase, no further hyperparameter tuning was performed. Instead, the hyperparameters of Baseline model 1 were used because it reached the best macro F1-Score during the validation stage. The extended models were trained using the same hyperparameters but the EfficientNet-B0, EfficientNet-B1 and EfficientNet-B2 classification architectures.

Finally, unweighted averaging was implemented to create an ensemble of the three models. The average ensemble was used to generate the final predictions.

4 Results

Results achieved for the different workflow stages are described in the subsequent sections. Classification results are summarized for the baseline models, the extended models, and their average ensembles. Additionally, the synthetic images generated for the extended training dataset are presented.

4.1 Baseline Model and Ensemble Performance

The classification results reached during an internal 5-fold CV are summarized in Table 5 and the classification results achieved for the test set are summarized in Table 6. The best macro F1-Score for a baseline model during CV was 92.11% ± 1.35 for baseline model 2. This model was an EfficientNet-B0 model trained with oversampling and the extended augmentation pipeline and was thus intended to generate more robust predictions. This model reached the best `infection` F1-Score of 60.25% for the test dataset. The best test F1-Score of 72.92% for the `none` class comparing the baseline models was reached for baseline model 4. This model was an EfficientNet-B1 model trained with a warm-up phase and the RMSprop [14] optimizer. This model reached a macro F1-Score of 56.40% and outperformed the remaining baseline models. Considering baseline models, the best test F1-Score for the `ischaemia` class was 47.50% reached for Baseline model 3. In comparison to the remaining baseline models, this model achieved the best F1-Score of 48.58% for the `both` class. Baseline model 3 was an EfficientNet-B2 model.

The average baseline ensemble reached a CV macro F1-Score of 90.37%±1.23. This result was slightly worse than the score of baseline model 2. For the test dataset, the average baseline ensemble outperformed all individual models for the F1-Score of the `none`, `ischaemia` and `both` classes, as well as for the macro F1-Score. The macro F1-Score of this model was 59.36%.

Table 5. Internal 5-fold CV classification results: Macro and class F1-Scores (F1). All scores are given as $\bar{x} \pm \sigma$. Best results are highlighted.

Model	`none` F1 %	`infection` F1 %	`ischaemia` F1 %	`both` F1 %	macro F1 %
B_1	86.31 ± 0.21	84.70 ± 0.96	82.18 ± 2.58	88.36 ± 2.56	85.39 ± 1.39
B_2	**90.61 ± 1.36**	**90.12 ± 1.30**	**91.92 ± 3.23**	**95.78 ± 1.47**	**92.11 ± 1.35**
B_3	80.38 ± 1.27	76.24 ± 1.46	67.83 ± 7.06	79.54 ± 1.56	76.00 ± 2.02
B_4	85.96 ± 0.48	84.15 ± 1.06	83.10 ± 2.86	89.57 ± 0.92	85.69 ± 0.81
$B_{ensemble}$	89.50 ± 0.39	88.41 ± 0.70	89.89 ± 4.66	93.70 ± 0.78	90.37 ± 1.23
E_1	85.64 ± 1.10	83.27 ± 1.45	82.84 ± 3.52	87.23 ± 2.58	84.75 ± 1.18
E_2	86.84 ± 0.59	84.28 ± 0.55	84.49 ± 2.82	89.01 ± 0.54	86.15 ± 0.61
E_3	88.20 ± 0.57	85.89 ± 0.85	86.42 ± 3.87	90.42 ± 1.45	87.73 ± 1.51
$E_{ensemble}$	89.15 ± 0.63	87.28 ± 0.73	90.12 ± 3.82	92.70 ± 1.03	89.81 ± 0.97

4.2 Synthetic Images for Training Dataset Extension

Generated synthetic images showed a broad variety regarding their quality and visual coherence, examples are shown in Fig. 4. While no realistically looking extremity-like structures such as toes were generated, contents usually resembled less or more convincing isolated ulcerated/ischaemic areas.

Qualitatively good and convincing results incorporated photo-realistic fine details, e.g., depth through multiple layers of skin with scale-like structures (Fig. 4a), granulation-like textures with wetness and reflection artifacts (Fig. 4b), infection-like localized redness (Fig. 4b and Fig. 4d), and localized cyanotic respectively necrotic coloring and textures (Fig. 4c and Fig. 4d). Qualitatively poor results suffered from either unsharp (Fig. 4e and Fig. 4f) or unconvincing (Fig. 4g and Fig. 4h) representations. The `ischaemia` and `both` models, in particular, trained with few images, were prone to generate less convincing synthetic images, compared to that generated by the `none` and `infection` models.

Table 6. Official classification results for the test part of the dataset: Macro, weighted average (WA), and class F1-Scores (F1) as well as the Accuracy. Best results are highlighted.

Model	`none` F1 %	`infection` F1 %	`ischaemia` F1 %	`both` F1 %	Acc. %	WA F1 %	macro F1 %
B_1	71.14	57.38	41.50	44.54	62.36	61.73	53.64
B_2	72.39	**60.25**	42.49	46.06	64.11	63.70	55.30
B_3	72.86	55.08	47.50	48.58	63.38	62.05	56.00
B_4	72.92	59.72	46.81	46.15	64.70	63.88	56.40
$B_{ensemble}$	74.24	59.54	51.67	51.97	66.08	65.06	59.36
E_1	74.36	58.46	54.02	51.22	65.99	64.66	59.51
E_2	74.09	59.15	55.49	50.56	66.01	64.85	59.82
E_3	74.41	59.05	54.30	53.32	66.34	65.13	60.27
$E_{ensemble}$ (2nd)	**74.53**	59.17	**55.80**	**53.59**	**66.57**	**65.32**	**60.77**

Generated color schemes were usually consistent, yet the `ischaemia` model tended to include blue areas (Fig. 4g), learned from occasional blue backgrounds in the few baseline images of the respective class.

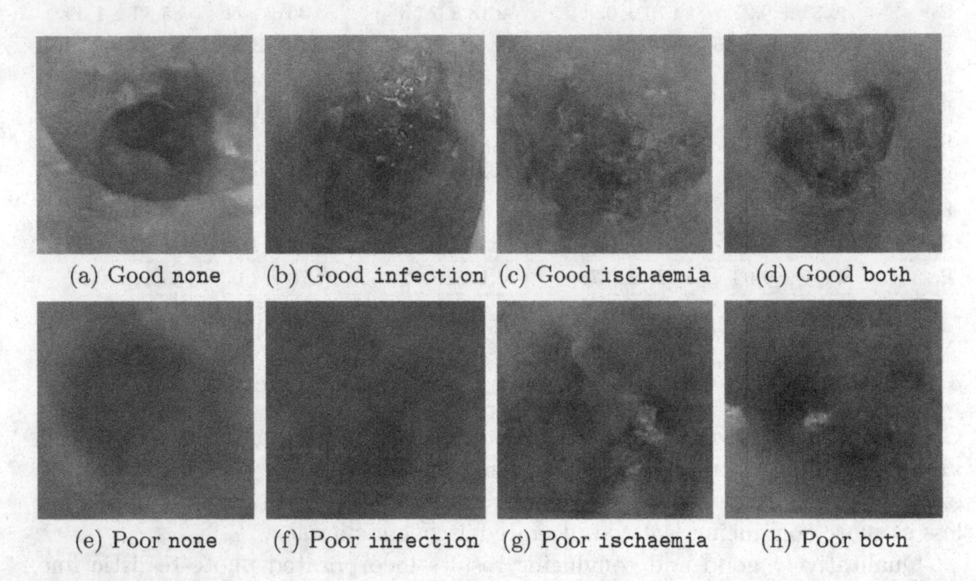

(a) Good `none` (b) Good `infection` (c) Good `ischaemia` (d) Good `both`

(e) Poor `none` (f) Poor `infection` (g) Poor `ischaemia` (h) Poor `both`

Fig. 4. Examples for generated synthetic images: (a)–(d) show qualitatively good, (e)–(h) qualitatively poor results. (Color figure online)

4.3 Extended Model and Ensemble Performances

Using the synthetic images, models were trained to improve the results of the average baseline ensemble. Due to time limitations during the challenge, the classification pipelines for these models were not as diverse as the baseline models. However, the three models differed in using multiple scales of the EfficientNet family. Table 5 summarizes the internal results during 5-fold CV and Table 6 summarizes the official results for the test set. The best macro F1-Score of 87.73% $\pm$ 1.51 during CV was reached by the model E_3, which used the EfficientNet-B2 architecture. This model reached the best test F1-Score for the `none` and `both` classes. As well as a test macro F1-Score of 60.27% that outperformed the remaining extended models. The best F1-Scores for the `infection` and `ischaemia` classes were achieved for model E_2, which was an EfficientNet-B1 model. The F1-Score was 59.15% for the `infection` class and 55.49% for the `ischaemia` class. All extended models outperformed the average baseline ensemble for the macro F1-Score. Increased F1-Scores can be especially noted for the `ischaemia` class.

The average ensemble outperformed the individual models for the macro F1-Score during CV, the test F1-Score for all classes, as well as for the test macro F1-Score. The macro F1-Score for the average extended ensemble was 60.77%.

The results of the three best placements are summarized in Table 7, the described ensemble model achieved 2nd place. This model outperformed the remaining models for the F1-score of the `ischaemia` class. More precise documentation about the challenge results are summarized in [4].

Table 7. Official classification results for the test part of the dataset and the three best challenge participants: Macro, weighted average (WA), and class F1-Scores (F1) as well as the Accuracy. Best results are highlighted.

Challenge placement	`none` F1 %	`infection` F1 %	`ischaemia` F1 %	`both` F1 %	Acc. %	WA F1 %	macro F1 %
1st place [10]	**75.74**	63.88	52.82	**56.19**	**68.56**	**68.01**	**62.16**
2nd place (this work)	74.53	59.17	**55.80**	53.59	66.57	65.32	60.77
3rd place	71.57	**67.14**	45.74	53.90	67.11	67.14	59.59

4.4 Local Interpretable Model-Agnostic Explanations (LIME)

Local Interpretable Model-agnostic Explanations (LIME)[12] [23] version `0.2.0.1` are used to visualize the model explanations of example images from the DFUC dataset provided by the maintainers. Per image 3,000 samples were generated to identify the most important superpixels. The 10 most important superpixels for each image are visualized in Fig. 5, predictions are summarized in Table 8.

Superpixels highlighted in green increase the probability of the predicted class (one vs. rest), whereas superpixels highlighted in red decrease the model probability of the predicted class.

Baseline and extended models as well as their ensembles do not tend to strongly focus on clinically non-relevant areas, such as backgrounds visible in example images I_5–I_8 for the classes `ischaemia` and `both`. Extended models as well as their ensemble also tend to be more certain regarding their predictions, involving greater probability-increasing superpixel areas as can be seen in true-positive predictions for I_2, I_3, and I_7. Yet, this also accounts for false-positive predictions for I_4, I_5, and I_8.

[12] LIME: https://github.com/marcotcr/lime, access 2021-11-12.

Fig. 5. Explainability: (a)–(h) show example images for the classes `none` (I_1, I_2), `infection` (I_3, I_4), `ischaemia` (I_5, I_6), and `both` (I_7, I_8) in the first row. The following five rows show LIME decision maps for the four baseline models B_1, B_2, B_3, B_4, and their ensemble B_{ensemble}. The last four rows show respective activation maps for the three extended models E_1, E_2, E_3, and their ensemble E_{ensemble}. Corresponding class predictions with confidences are listed in Table 8. (Color figure online)

Table 8. Summary of class predictions of models B_1–B_4, $B_{ensemble}$, E_1–E_3, and $E_{ensemble}$ for example images I_1–I_8, shown with LIME decision maps in Fig. 5. False-positive class predictions are highlighted in red.

Model	I_1 class conf. % (none)	I_2 class conf. % (none)	I_3 class conf. % (inf.)	I_4 class conf. % (inf.)	I_5 class conf. % (isc.)	I_6 class conf. % (isc.)	I_7 class conf. % (both)	I_8 class conf. % (both)
B_1	inf. 61.02	none 99.99	inf. 99.01	none 99.14	both 98.12	isc. 63.06	isc. 83.53	inf. 99.75
B_2	none 89.55	none 99.74	inf. 99.99	inf. 99.98	isc. 99.98	both 88.32	isc. 99.99	inf. 99.88
B_3	inf. 81.64	none 99.83	none 78.51	none 92.37	isc. 61.49	isc. 92.83	inf. 88.43	inf. 98.57
B_4	inf. 90.30	none 84.10	inf. 97.43	none 96.58	both 87.66	isc. 85.47	inf. 80.20	inf. 72.77
$B_{ensemble}$	inf. 60.85	none 95.92	inf. 78.81	none 72.03	both 55.01	isc. 63.26	isc. 47.05	inf. 92.74
E_1	none 91.95	none 100.00	inf. 100.00	none 99.99	both 91.30	isc. 99.92	both 97.27	inf. 100.00
E_2	none 99.87	none 100.00	inf. 100.00	none 100.00	both 99.97	isc. 88.33	inf. 94.54	inf. 100.00
E_3	none 98.66	none 100.00	inf. 100.00	none 100.00	both 99.97	isc. 100.00	both 99.64	inf. 99.99
$E_{ensemble}$	none 96.83	none 100.00	inf. 100.00	none 100.00	both 97.08	isc. 96.08	both 67.46	inf. 100.00

5 Discussion

In the following, experiments and results including the model and ensemble development as well as pseudo-labeling and synthetic image generation are discussed. Limitations of the work and in the experiment design are addressed.

5.1 Models and Ensembles

In this research, state of the art deep learning-based models were used for DFU infection and ischaemia classification. During the validation stage, explorative investigations with different deep-learning architectures and hyperparameters were performed. The best validation F1-Scores for all individual classes were achieved for EfficientNets. More complex models like EfficientNet-v2 and Vision Transformers achieved worse results than the EfficientNets during the validation stage of the challenge. Those best-performing models were combined to an ensemble that outperformed the individual models for the test F1-Score of the `none`, `ischaemia` and `both` classes and for the macro F1-Score.

The models trained on the extended training dataset outperformed those trained on the baseline training dataset. Extended models achieved outstanding results for the `ischaemia` class. The averaged extended model ensemble reached the overall best macro F1-Score of 60.77% and the best class F1-Scores for the `none`, `ischaemia`, and `both` classes. Training on the extended training dataset, therefore, led to considerable improvements for the rare classes `ischaemia` and `both`, without harming classification performance for the common classes `none` and `infection`. The best model of the challenge benchmark experiments [33] reached an F1-Score of 55.00%. The research at hand outperformed this result by 10.49% (5.77% points). The explanations generated via LIME do not indicate that models and ensembles strongly focus on medically irrelevant areas such as backgrounds.

5.2 Pseudo-Labeling and Synthetic Image Generation

Pseudo-labeling of unlabeled data, either for regular dataset extension or self-training approaches, is usually a practicable way to extend available training data, fostering generalization of models. This technique already proved beneficial for a detection task on DFUs [34]. In the presented work, creation of pseudo-labeled images allowed to notably increasing the amount of available training data, especially for the rare classes `ischaemia` and `both`. Beside images of the test part of the dataset, unlabeled images of the training part were a viable source. The chosen high confidence threshold for inclusion of pseudo-labels is assumed to withheld ingress of the majority of misclassifications, however, no further investigation on this matter was conducted.

The achieved increase of available training images was crucial for class-individual pix2pixHD model training, in particular for the classes `ischaemia` and `both`. During initial experiments on the original training part, randomly chosen results of models for these classes solely displayed unconvincing results of poor quality and lacking detail. After the pseudo-label extension, randomly chosen results of re-trained models showed a notably increased quality and higher level of details. Results of the `none` and `infection` class models benefited as well, yet initial results usually displayed sufficient detail. The broad variety of results generated by final pix2pixHD models traces back to that of the DFUC 2021 dataset. Missing representations of realistically looking extremities attribute to the overall majority of training images that show small DFUs, solely surrounded by skin. Qualitatively poor and detail-lacking results of `none` and `infection` class models were convincing to this extent, that they may be associated with images resulting from poor imaging. However, those of the `ischaemia` and `both` class models partially featured unnatural coloring and repeating patterns. This indicates, that even though generation of adequate results is achievable with a few hundred training images, at least a few thousand are required to achieve consistently convincing results with pix2pixHD.

The aggressive approach of class imbalance compensation with massive amounts of synthetic images essentially improved the extended EfficientNets model ensemble performance for the rare classes `ischaemia` and `both`. In contrast, performance for the common classes `none` and `infection` did not suffer despite considerable amounts of qualitatively poor and potentially unconvincing samples were part of the overall extension. As color schemes of synthetic images were consistent regardless of their quality, beneficial effects may be rather attributable to these than to fine details of synthesized patterns.

5.3 Limitations

The approach proposed in this article features several limitations. First, during the validation stage, only explorative experiments were performed. However, to get a better insight into which deep-learning models, augmentation pipelines and hyperparameters performed best a structural comparison is important. A structural comparison can for example include grid-search or ablation studies. Future work should include the investigation of more recent deep learning models, e.g., Residual Convolutional Neural Split-Attention Network (ResNeSt) [38], Class-Attention in Image Transformers (CaiT) [30] or Data-efficient Image Transformers (DeiT) [29]. This also applies to the experiments on the extended dataset. Due to time limitations during the validation stage of the challenge, no validation experiments were executed to identify the best-performing models for each class trained on the extended dataset. An attempt to clear the training dataset from augmented images using Scale-Invariant Feature Transform (SIFT) [18] in order to achieve unbiased CV results was not successful. Future work should investigate further dataset cleansing strategies to address this problem. However, the exclusion of too similar images in the original datasets via hashing impedes the cleansing. More sophisticated ensembling strategies can further improve classification results.

Regarding the presented training dataset extension strategy both, the pseudo-labeling and synthetic image generation approach can be optimized. The threshold for inclusion of pseudo-label candidates was chosen conservatively on purpose and not evaluated via validation experiments. Hence, a more balanced choice is possible to achieve a greater or more qualitative outcome of additional training images. Consequently, these have a direct influence on the pix2pixHD models and extended EfficientNets model ensemble. For generated synthetic images via pix2pixHD models no metric-based analysis or quality assessment was conducted, hence there was no filtering of potentially harming samples. In addition, the visual assessment of these images was performed by non-clinicians. Hence, no statements on the actual convincibility regarding realistic looks in the eyes of clinicians can be made. Further, the applied class-balancing approach relying on massive amounts of synthetic images was rather aggressive. A more subtle approach with less extension for classes with an already sufficient amount of training images might enable better overall performance. In addition, unconditional GANs may have displayed a better choice for the given classification task as these do not require masks for training or generation. Respective recent developments such as StyleGAN2+ADA[13] [16] further enable data efficiency via adaptive discriminator augmentation, facilitating qualitative results for rather small amounts of training images.

6 Conclusion

This work investigated, whether training dataset extension with pseudo-labels and synthetic images generated by pix2pixHD can improve EfficientNet-based model ensemble performance for infection and ischaemia classification in DFUs. For evaluation, the amount of 5,955 labeled images of the training part of the DFUC 2021 dataset was extended with (i) 6,961 pseudo-labeled images from unlabeled images in the training and test part, and (ii) 38,748 synthetic images for subsequent class-balancing. The resulting extended training part had 8.67 times the size of the baseline training dataset with a real to synthetic image ratio of 1:3, featuring manifolds of synthetic images for the rare classes `ischaemia` and `both`.

Results show that the macro F1-Scores of the averaged baseline model ensembles outperformed the individual classifiers. All models trained on the extended dataset outperformed the baseline ensemble for the macro F1-Score. In particular, considerable improvements of the class F1-Scores for rare classes were achieved while no harming effects for common classes were detected. The best results were achieved for the averaged extended model ensemble.

Pseudo-labeling represents an effective strategy to extend datasets. Extension and class balancing via synthetic images generated by GANs has the potential to further improve the overall performance of classification models, especially that for rare classes, given a sufficient amount of images for training.

Acknowledgments. Louise Bloch and Raphael Brüngel were partially funded by PhD grants from University of Applied Sciences and Arts Dortmund, Dortmund, Germany. The authors thank Henryk Birkhölzer for advice on pix2pixHD.

References

1. Alzubaidi, L., Fadhel, M.A., Oleiwi, S.R., Al-Shamma, O., Zhang, J.: DFU_QUTNet: diabetic foot ulcer classification using novel deep convolutional neural network. Multimed. Tools Appl. **79**(21), 15655–15677 (2019). https://doi.org/10.1007/s11042-019-07820-w
2. Buslaev, A., Iglovikov, V.I., Khvedchenya, E., Parinov, A., Druzhinin, M., Kalinin, A.A.: Albumentations: fast and flexible image augmentations. Information **11**(2), 125 (2020). https://doi.org/10.3390/info11020125
3. Canny, J.: A computational approach to edge detection. IEEE Trans. Pattern Anal. Mach. Intell. **8**(6), 679–698 (1986). https://doi.org/10.1109/tpami.1986.4767851
4. Cassidy, B., et al.: Diabetic foot ulcer grand challenge 2021: evaluation and summary. arXiv preprint arXiv:2111.10376 (2021)

[13] StyleGAN2+ADA: https://github.com/NVlabs/stylegan2-ada, access 2021-09-22.

5. Das, S.K., Roy, P., Mishra, A.K.: DFU_SPNet: a stacked parallel convolution layers based CNN to improve diabetic foot ulcer classification. ICT Express (2021). https://doi.org/10.1016/j.icte.2021.08.022
6. Das, S.K., Roy, P., Mishra, A.K.: Recognition of ischaemia and infection in diabetic foot ulcer: a deep convolutional neural network based approach. Int. J. Imaging Syst. Technol. (2021). https://doi.org/10.1002/ima.22598
7. Deng, J., Dong, W., Socher, R., Li, L., Li, K., Fei-Fei, L.: ImageNet: a large-scale hierarchical image database. In: Proceedings of the IEEE Conference on Computer Vision and Pattern Recognition (CVPR 2009), pp. 248–255. IEEE (2009). https://doi.org/10.1109/cvpr.2009.5206848
8. Dosovitskiy, A., et al.: An image is worth 16x16 words: transformers for image recognition at scale. In: Proceedings of the 9th International Conference on Learning Representations (ICLR 2021) (2021)
9. Falanga, V.: Wound healing and its impairment in the diabetic foot. The Lancet **366**(9498), 1736–1743 (2005). https://doi.org/10.1016/s0140-6736(05)67700-8
10. Galdran, A., Carneiro, G., Ballester, M.A.G.: Convolutional nets versus vision transformers for diabetic foot ulcer classification. arXiv preprint arXiv:2111.06894 (2021)
11. Goodfellow, I., et al.: Generative adversarial nets. In: Ghahramani, Z., Welling, M., Cortes, C., Lawrence, N., Weinberger, K.Q. (eds.) Advances in Neural Information Processing Systems (NIPS 2017), vol. 27. Curran Associates, Inc. (2014)
12. Goyal, M., Reeves, N.D., Davison, A.K., Rajbhandari, S., Spragg, J., Yap, M.H.: DFUNet: convolutional neural networks for diabetic foot ulcer classification. IEEE Trans. Emerg. Top. Comput. Intell. **4**(5), 728–739 (2020). https://doi.org/10.1109/tetci.2018.2866254
13. He, K., Zhang, X., Ren, S., Sun, J.: Deep residual learning for image recognition. In: Proceedings of the IEEE Conference on Computer Vision and Pattern Recognition (CVPR 2016), pp. 770–778 (2016). https://doi.org/10.1109/CVPR.2016.90
14. Hinton, G., Srivastava, N., Swersky, K.: Lecture 6e rmsprop: divide the gradient by a running average of its recent magnitude (2012). https://www.cs.toronto.edu/~tijmen/csc321/slides/lecture_slides_lec6.pdf
15. Isola, P., Zhu, J.Y., Zhou, T., Efros, A.A.: Image-to-image translation with conditional adversarial networks. In: Proceedings of the IEEE Conference on Computer Vision and Pattern Recognition (CVPR 2017). IEEE (2017). https://doi.org/10.1109/cvpr.2017.632
16. Karras, T., Aittala, M., Hellsten, J., Laine, S., Lehtinen, J., Aila, T.: Training generative adversarial networks with limited data. In: Larochelle, H., Ranzato, M., Hadsell, R., Balcan, M.F., Lin, H. (eds.) Advances in Neural Information Processing Systems (NeurIPS 2020), vol. 33, pp. 12104–12114. Curran Associates, Inc. (2020)
17. Krizhevsky, A., Sutskever, I., Hinton, G.E.: ImageNet classification with deep convolutional neural networks. In: Pereira, F., Burges, C.J.C., Bottou, L., Weinberger, K.Q. (eds.) Advances in Neural Information Processing Systems (NIPS 2012), vol. 25. Curran Associates, Inc. (2012)
18. Lowe, D.G.: Distinctive image features from scale-invariant keypoints. Int. J. Comput. Vis. **60**(2), 91–110 (2004). https://doi.org/10.1023/b:visi.0000029664.99615.94
19. Merkel, D.: Docker: lightweight Linux containers for consistent development and deployment. Linux J. **2014**(239), 2 (2014)
20. Micikevicius, P., et al.: Mixed precision training. In: Proceedings of the 6th International Conference on Learning Representations (ICLR 2018) (2018)

21. Mirza, M., Osindero, S.: Conditional generative adversarial nets. arXiv preprint arXiv:1411.1784 (2014)
22. Paszke, A., et al.: PyTorch: an imperative style, high-performance deep learning library. In: Wallach, H., Larochelle, H., Beygelzimer, A., d'Alché Buc, F., Fox, E., Garnett, R. (eds.) Advances in Neural Information Processing Systems (NeuriPS 2019), vol. 32, pp. 8024–8035. Curran Associates, Inc. (2019)
23. Ribeiro, M.T., Singh, S., Guestrin, C.: Why should I trust you? Explaining the predictions of any classifier. In: Proceedings of the 22nd ACM International Conference on Knowledge Discovery and Data Mining (SIGKDD 2016), pp. 1135–1144 (2016). https://doi.org/10.1145/2939672.2939778
24. Saeedi, P., et al.: Global and regional diabetes prevalence estimates for 2019 and projections for 2030 and 2045: results from the international diabetes federation diabetes atlas, 9th edition. Diabetes Res. Clin. Pract. **157**, 107843 (2019). https://doi.org/10.1016/j.diabres.2019.107843
25. Sarp, S., Kuzlu, M., Wilson, E., Guler, O.: WG^{2}AN: synthetic wound image generation using generative adversarial network. J. Eng. **2021**(5), 286–294 (2021). https://doi.org/10.1049/tje2.12033
26. Siddiqui, A.R., Bernstein, J.M.: Chronic wound infection: facts and controversies. Clin. Dermatol. **28**(5), 519–526 (2010). https://doi.org/10.1016/j.clindermatol.2010.03.009
27. Tan, M., Le, Q.: EfficientNet: rethinking model scaling for convolutional neural networks. In: Chaudhuri, K., Salakhutdinov, R. (eds.) Proceedings of the 36th International Conference on Machine Learning (ICML). Proceedings of Machine Learning Research (PMLR 2019), vol. 97, pp. 6105–6114. PMLR (2019)
28. Tan, M., Le, Q.: EfficientNet: rethinking model scaling for convolutional neural networks. In: K. Chaudhuri, R. Salakhutdinov (eds.) Proceedings of the 36th International Conference on Machine Learning (ICML), Proceedings of Machine Learning Research (PLMR 2019), vol. 97, pp. 6105–6114. PMLR (2019)
29. Touvron, H., Cord, M., Douze, M., Massa, F., Sablayrolles, A., Jegou, H.: Training data-efficient image transformers distillation through attention. In: Proceedings of the International Conference on Machine Learning (ICML 2021), vol. 139, pp. 10347–10357 (2021)
30. Touvron, H., Cord, M., Sablayrolles, A., Synnaeve, G., Jégou, H.: Going deeper with image transformers. In: Proceedings of the IEEE/CVF International Conference on Computer Vision (ICCV 2021), pp. 32–42 (2021)
31. Wang, T.C., Liu, M.Y., Zhu, J.Y., Tao, A., Kautz, J., Catanzaro, B.: High-resolution image synthesis and semantic manipulation with conditional GANs. In: Proceedings of the IEEE/CVF Conference on Computer Vision and Pattern Recognition (CVPR 2018). IEEE (2018). https://doi.org/10.1109/cvpr.2018.00917
32. Wightman, R.: PyTorch image models (2019). https://doi.org/10.5281/zenodo.4414861. https://github.com/rwightman/pytorch-image-models
33. Yap, M.H., Cassidy, B., Pappachan, J.M., O'Shea, C., Gillespie, D., Reeves, N.D.: Analysis towards classification of infection and ischaemia of diabetic foot ulcers. In: Proceedings of the IEEE EMBS International Conference on Biomedical and Health Informatics (BHI 2021), pp. 1–4 (2021). https://doi.org/10.1109/BHI50953.2021.9508563
34. Yap, M.H., et al.: Deep learning in diabetic foot ulcers detection: a comprehensive evaluation. Comput. Biol. Med. **135**, 104596 (2021). https://doi.org/10.1016/j.compbiomed.2021.104596
35. Yap, M.H., et al.: Diabetic Foot Ulcers Grand Challenge 2020. https://doi.org/10.5281/zenodo.3731068

36. Yap, M.H., et al.: Diabetic Foot Ulcers Grand Challenge 2021. https://doi.org/10.5281/zenodo.3715020
37. Yap, M.H., et al.: Diabetic Foot Ulcers Grand Challenge 2022. https://doi.org/10.5281/zenodo.4575228
38. Zhang, H., et al.: ResNeSt: split-attention networks. arXiv preprint arXiv:2004.08955 (2020)
39. Zhang, J., Zhu, E., Guo, X., Chen, H., Yin, J.: Chronic wounds image generator based on deep convolutional generative adversarial networks. In: Li, L., Lu, P., He, K. (eds.) Theoretical Computer Science. NCTCS 2018. CCIS, vol. 882, pp. 150–158. Springer, Singapore (2018). https://doi.org/10.1007/978-981-13-2712-4_11
40. Zhang, P., Lu, J., Jing, Y., Tang, S., Zhu, D., Bi, Y.: Global epidemiology of diabetic foot ulceration: a systematic review and meta-analysis. Ann. Med. **49**(2), 106–116 (2016). https://doi.org/10.1080/07853890.2016.1231932

Bias Adjustable Activation Network for Imbalanced Data—Diabetic Foot Ulcer Challenge 2021

Salman Ahmed(✉) and Hammad Naveed

National University of Computer and Emerging Sciences, Islamabad, Pakistan
s.ahmed@nu.edu.pk

Abstract. Despite great success, deep learning models still face a critical obstacle in classifying highly imbalanced real life data. Detection of diabetic foot ulcer is fundamental for healthcare specialists to prevent amputations. In this work, we performed multiple experiments to benchmark results on Grand Challenge for Diabetic Foot Ulcer Detection 2021. To adjust the bias of the convolutional neural networks, we also propose a custom designed activation layer based on softmax to handle the probability skew of the classes. We achieved 2nd position in the validation set with a Macro F1-Score of 0.593 and 3rd position in Test set with a Macro F1-Score of 0.596 of the Diabetic Foot Ulcer Detection 2021 Grand Challenge.

Keywords: Medical imaging · Deep learning · Diabetic Foot Ulcer

1 Introduction

Diabetes cases are increasing at an alarming rate across the world. In 2005, Center of Disease Control and Prevention approximated that 20.8 million people, roughly 7% of United States population had diabetes. In 2005 alone, 1.5 million new cases of diabetes were diagnosed in people aged 20 years or older [1].

In 2020, Center of Disease Control and Prevention reported that 34.2 million Americans had diabetes and 88 million adults had pre-diabetes. Newly diagnosed cases of diabetes significantly increased among US youth in 2020. Diabetic foot ulceration affects approximately 56% of diabetic patients and 20% of foot ulcer patients end up with some type of amputation [1].

Several approaches based on Artificial Intelligence and Machine Learning (ML) have been used for Diabetic Foot Ulcers detection [2]. Multiple feature extraction techniques are used to extract features and SVM classifier is used as traditional ML approach. Many of these feature extraction methods are based on texture and color descriptions of the image [3]. These descriptors can include effected temperature and light condition of the image. In contrast deep learning methods do not require manual feature extraction as this process is done

M. H. Yap et al. (Eds.): DFUC 2021, LNCS 13183, pp. 50–61, 2022.
https://doi.org/10.1007/978-3-030-94907-5_4

automatically to extract feature on high non linear dimensions in the training process [4].

Diabetic Foot Ulcers (DFU) are a serious complication of diabetes affecting one in three people with diabetes [5]. Diabetic Foot Ulcers Grand Challenge (DFUC) 2021 is a collaboration of lead scientists from UK, US, India and New Zealand [5]. The DFUC 2021 dataset consists of foot ulcer images collected from Lancashire Teaching Hospital over the past few years. Kodak DX4530, Nikon D3300 and Nikon COOLPIX P100 cameras were used to capture foot images in room lights (instead of flash) [5,6].

The benchmark performance of five key backbones of deep learning, i.e. VGG16 [7], ResNet101 [8], InceptionV3 [9], DenseNet121 [10] and EfficientNet [11] on DFUC2021 dataset was reported by Yap et al. [5]. EfficientNet-B0 backbone obtained the best Macro F1-Score of 0.55 by using data augmentation and transfer learning. Details of the DFUC2021 challenge can be found on the following link: https://dfu-challenge.github.io.

We analyzed and reproduced these experiments by Yap et al. [5]. Moreover, we explored multiple convolutional and transformer based architectures to understand the best way to solve this problem. It was demonstrated that machine learning methods have short comings when analyzing imbalanced datasets [12]. For example, if a disease is in 1% patients of the dataset then a model that never predicts such a disease is 99% accurate. Alternative measures of good-fit are F1-Score, precision and recall. Previous studies introduced multiple heuristics like weighted-labels, label-smoothing, over-sampling and under-sampling [13] to solve this problem.

In this paper, we introduce a new activation layer "Bias Adjustable Softmax" for imbalanced datasets, which is a modification in standard softmax function that uses class weights to adjust skew of biases in the architecture for highly imbalanced datasets. Furthermore, we train, compare, and analyze state-of-the-art ImageNet architectures with Bias Adjustable Softmax against standard Softmax on DFUC 2021 dataset.

2 Methodology

DFUC2021 challenge dataset consists of 15683 images with 5955 images in the training set (2,555 infection only, 227 ischaemia only, 621 both infection and ischaemia, and 2,552 without ischaemia and infection), 3994 unlabeled images and 5734 images in the testing set. We used the same data split as DFUC 2021 [5]. The complete dataset is available at https://dfu challenge.github.io.

The performance metrics used in the DFUC 2021 challenge are Macro-F1, Area under the curve, Macro-precision, Macro-Recall. Macro-F1 is usually considered as a better metric to evaluate the model performance on highly imbalanced data. As the Leader-board of Grand Challenge DFUC 2021 is evaluated on Macro-F1 Score therefore, in this study we used Macro-F1 score as the primary evaluation metric in our experiments.

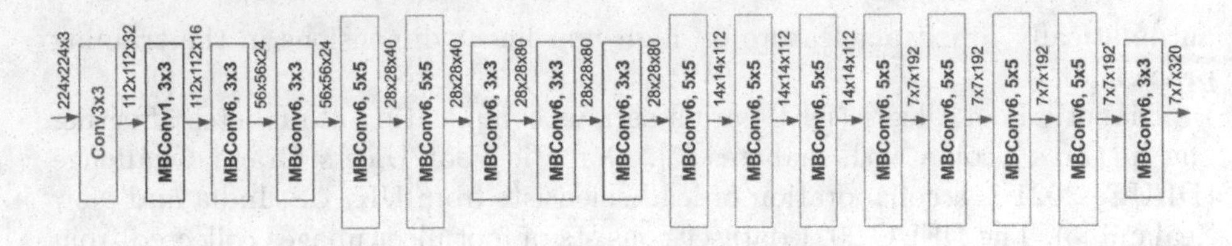

Fig. 1. EfficientNet B0 architecture

Experimental Setup. We used pytorch [14] as the deep learning framework to perform our experiments. In the benchmark study [5], EfficientNet-B0 performed better than other pre-trained convolutional architectures. Therefore, we performed our experiments with EfficientNet backbones and Resnet (a popular architecture for bench marking).

EfficientNet-B0 was designed by AutoML MNAS Framework by performing a neural architecture search which optimizes both accuracy and number of parameters [11]. EfficientNet B0 uses mobile inverted convolutions (MBConv), which were originally used in MobileNetV2 [15]. Figure 1 shows the architecture of EfficientNet B0.

2.1 Data Augmentation

As the dataset contained limited images for each class and the distribution of the images was highly imbalanced, it was important to use data augmentation to generalize the model. We used albumentations [12] with pytorch to add augmentations. We used the following augmentations: Horizontal Flip, Random Brightness, Random Contrast, Blur, Median Blur, Gaussian Blur, Motion Blur, Optical Distortion, Grid Distortion, Hue Saturation, and Shift Scale Rotate. Figure 2 shows a batch with and without these augmentations. We performed all experiments with the image size of 224×224.

2.2 Scheduler Adjustments

We used gradual warm-up that increase learning rate from small to a large value for some epochs. This increase in learning rate from small to large avoids a sudden jump which allows healthy convergence [16]. Figure 3 represents our case where we used gradual warm-up for 1 epoch which allows us to train our model with learning rate of 0.0001 for 1 epoch. After gradual warm-up we adjust our model to Cosine Annealing learning rate scheduler [17]. We used Cosine Annealing learning rate scheduler for 88 epochs. In our case Cosine Annealing learning rate scheduler starts after the first epoch and adjusts the learning rate to 0.001. It decreases the learning rate using Cosine Annealing in each epoch.

2.3 Weighted Cross Entropy

During the training phase of the neural network, the loss function plays an important role in adjusting neural network weights in a way that it minimizes the

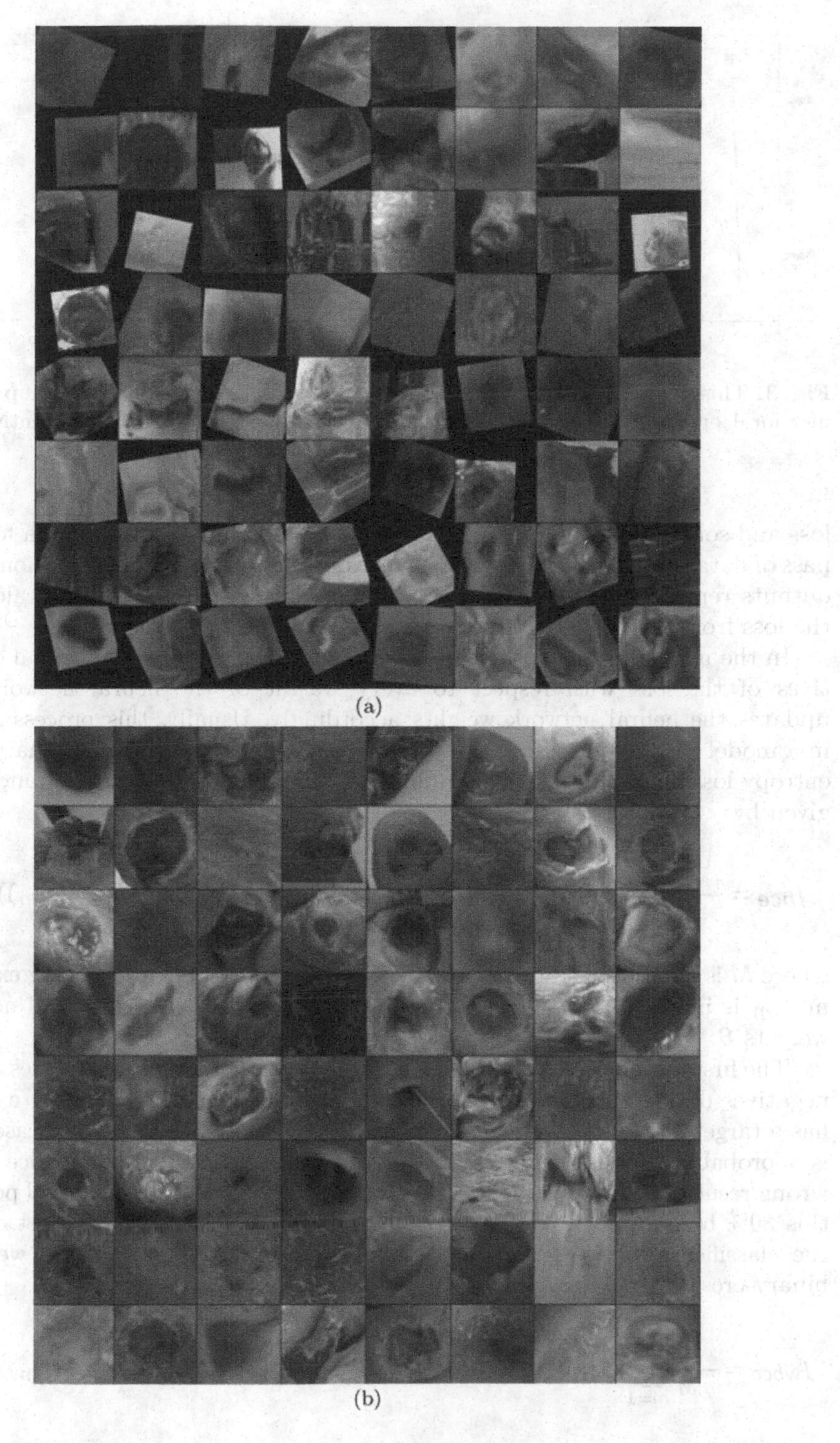

Fig. 2. Comparison of images with and without augmentations: (a) training batch with augmentations, (b) training batch with no augmentations

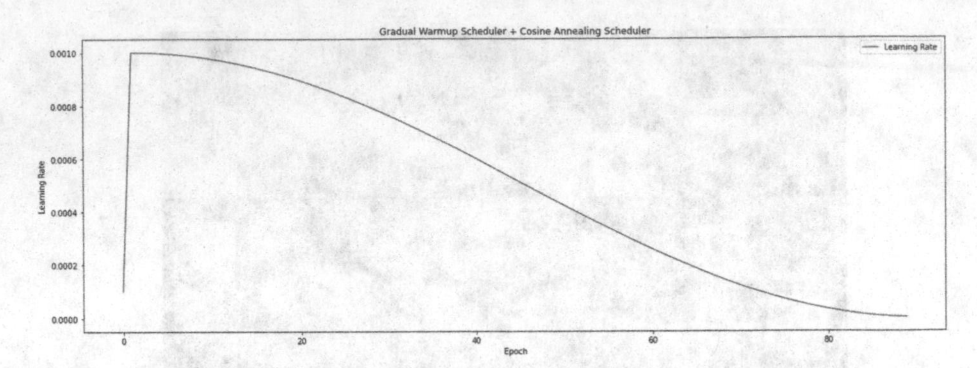

Fig. 3. This figure shows the learning rate adjustment using Gradual Warmup Scheduler for 1 epoch and then Cosine Annealing Scheduler for 88 epochs EfficientNet-B0.

loss and solves the given problem. The process of training starts with a forward pass of data through the model to generate outputs. In case of classification, these outputs represent probabilities of each class. The loss function then calculates the loss from the probabilities of each class and ground truths.

In the next step, the optimization function takes as input the partial derivatives of the loss with respect to every weight of the neural network and updates the neural network weights accordingly. Usually, this process results in a model with lower loss. In case of binary classification problem, binary-cross entropy loss function is used. The standard binary-cross entropy loss function is given by:

$$J_{bce} = -\frac{1}{M}\sum_{m=1}^{M}[y_m \times \log(h_\theta(x_m)] + [(1-y_m) \times \log(1-h_\theta(x_m))] \quad (1)$$

where M is number of training examples, y_m is target label for training example m, x_m is input for training example m, and h_θ is model with neural network weights θ.

The first term of the loss $y_m \times \log(h_\theta(x_m))$ restrains the probabilities of false negatives during the training phase. For example, consider a training example has a target label of 1, whereas the output of the model is 0.7. In this case there is a probabilistic false negative of 30%. The model has 30% confidence in the wrong result and 70% confidence is correct result. The loss function will penalize this 30% by a value of $-\log(0.7) = 0.155$. Now consider the perfect case, if the classifier returns 1 then the loss is $-\log(1) = 0$. The standard weighted binary-cross entropy loss function is given by:

$$J_{wbce} = -\frac{1}{M}\sum_{m=1}^{M}[w \times [y_m \times \log(h_\theta(x_m)] + [(1-y_m) \times \log(1-h_\theta(x_m))]] \quad (2)$$

where M is the number of training examples, w is the weight, y_m is the target label for the training example m, x_m is the input for training example m, and h_θ is the model with neural network weights θ. This additional weight w can be used to adjust the importance of positive labels. For the imbalanced classification more weight is given to the minority classes. The standard weighted categorical-cross entropy loss function is given by:

$$J_{wcce} = -\frac{1}{M} \sum_{k=1}^{K} \sum_{m=1}^{M} [w_k \times y_m^k \times \log(h_\theta(x_m, k))] \tag{3}$$

where M is the number of training examples, K is the number of classes, w is the weight of class k, y_m^k is the target label for training example m for class k, x_m is the input for training example m, and h_θ is the model with neural network weights θ. The most important term that modifies behavior of model with respect to each weight is $y_m^k \times \log(h_\theta(x_m, k))$. Multiplying this term by more weights for a class in cross entropy loss function will increase the confidence of model towards that class.

Weighted cross entropy loss function is perfect in theory. However, most models are trained in mini-batches. Each batch is chosen randomly from the training data during the training phase. Now consider if the minority classes occur in one or two mini-batches through out the epoch and every other mini-batch is completely based on the dominant classes. This will result in more confidence of models towards the dominant class(es).

Table 1. Class weights for categorical cross-entropy loss function

	Control	Infection	Ischaemia	Both
Number of images per class	2552	2555	227	621
Weights per class	0.58	0.55	8.33	2.84

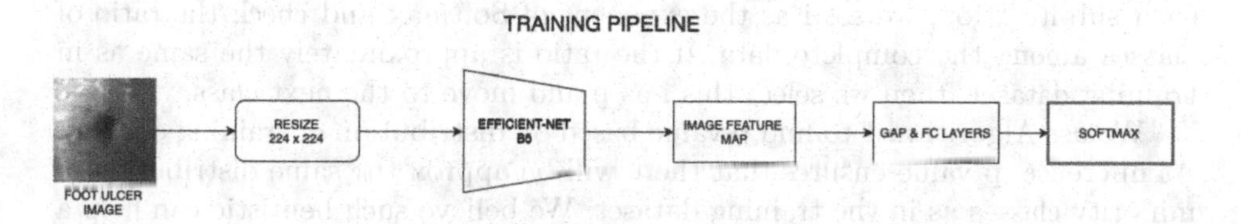

Fig. 4. This figure represents training pipeline used in each experiment.

EfficientNet B0-B6, Resnet-50, and Resnet-101 were fine-tuned on DFUC-2021 train dataset. We used weighted categorical cross-entropy to train these pipelines. Table 1 represents class weights used in categorical cross-entropy. Figure 4 represents training pipeline. We trained multiple models on DFUC 2021 challenge data with weighted categorical cross entropy but no significant impact on the results was seen on the highly imbalanced dataset.

2.4 Bias-Adjustable Softmax

We used class-weights in the loss function to handle class-imbalance. However, this resulted in our deep neural networks being biased towards Control and Infection classes because of highly imbalanced classes. Due to the highly imbalanced classes, deep neural networks naturally adjust their weights based on the dominant classes. The Softmax function can be represented as follows:

$$\text{Standard Softmax} = \frac{\exp(x_i)}{\sum_{j=1}^{J} \exp(x_j)} \tag{4}$$

where x is the input logits to softmax, J is the number of classes.

We introduced a novel way to adjust the skew of the probabilities for each class to adjust the bias at inference level. We modified the Standard Softmax by adjusting a parameter p.

$$\text{Bias Adjustable Softmax} = \left(\frac{\exp(x_i)}{\sum_{j=1}^{J} \exp(x_j)}\right)^{p} \tag{5}$$

where x is the input logits to softmax, J is the number of classes. The p value is dependent on class weights and its value adjusts the skew of class probabilities. We use Standard Softmax in Forward pass, and Back-propagation whereas, Bias-Adjustable Softmax is only used at inference and training pipeline remains the same.

Bias Adjustable Softmax Heuristic. A heuristic to find value of p for Softmax is shown in Algorithm 1. We start by initializing number of classes C and maximum search space M. We iterate for each class C to search for its p value. We sub-iterate from 0 to M with a step size of 0.01 to search p value as i. In each sub-iteration, we set i as the exponent of Softmax and check the ratio of classes among the complete data. If the ratio is approximately the same as in training dataset, then we select this i as p and move to the next class.

We use Algorithm 1 to find p value based on distribution of training dataset. At inference, p value ensures that there will be approx the same distribution of minority classes as in the training dataset. We believe such heuristic can help a model to adjust probability curve in case of highly imbalanced classes.

2.5 Inference Pipeline

Our inference pipeline is designed by combining the trained EfficientNet-B0 and Bias Adjustable Softmax. Bias Adjustable Heuristic converged with values 1.6,

Algorithm 1. Heuristic to find p

```
Initialize C                          ▷ C : Number of Classes
Initialize M                          ▷ M : Max search space (Default Value 5)
Initialize j = 0
while j < C do
    X = C_j                           ▷ X : Selected Class
    PD                                ▷ X : Probability Distribution of Selected Class
    TD                                ▷ X : Training Set Distribution of Selected Class
    Initialize i = 0
    while i < M do
        if PD^i ≈ TD then
            P_j = i
            break
        end if
        i = i + 0.1
    end while
    j = j + 1
end while
```

0.4, 2.2, and 0.55 respectively for classes Control, Infection, Ischaemia, and Both. Our Inference pipeline is shown in Fig. 5.

3 Results and Discussion

We compared the performance of Bias-Adjustable Softmax Classification with Standard Softmax training pipelines. We used the EfficientNet B0-B6, Resnet-50, and Resnet-101 as backbones that were pre-trained on ImageNet. All models were trained on Nvidia RTX 3090 for 90 Epochs with Adam optimizer and a batch size of 64.

To measure the effectiveness of Bias Adjustable Softmax, we selected 70% to 80% stratified subset randomly from the training set for each epoch. We calculated the class weights of the selected subset and trained on this subset. We trained for N number of epochs. The effect of p values based on a stratified subset of training distribution can be seen in Fig. 4.

Table 2 represents the comparison of Bias-Adjustable Softmax and Standard Softmax on Macro F1, and F1 Score for 4 classes (None, Infection, Ischaemia, and Both) in the family of EfficientNets and also represent the comparison of Bias-Adjustable Softmax and Standard Softmax on models from Benchmark study [5]. EfficientNet-B0 achieved best Validation and Test Macro-F1 of 0.59 for us on the DFUC challenge dataset.

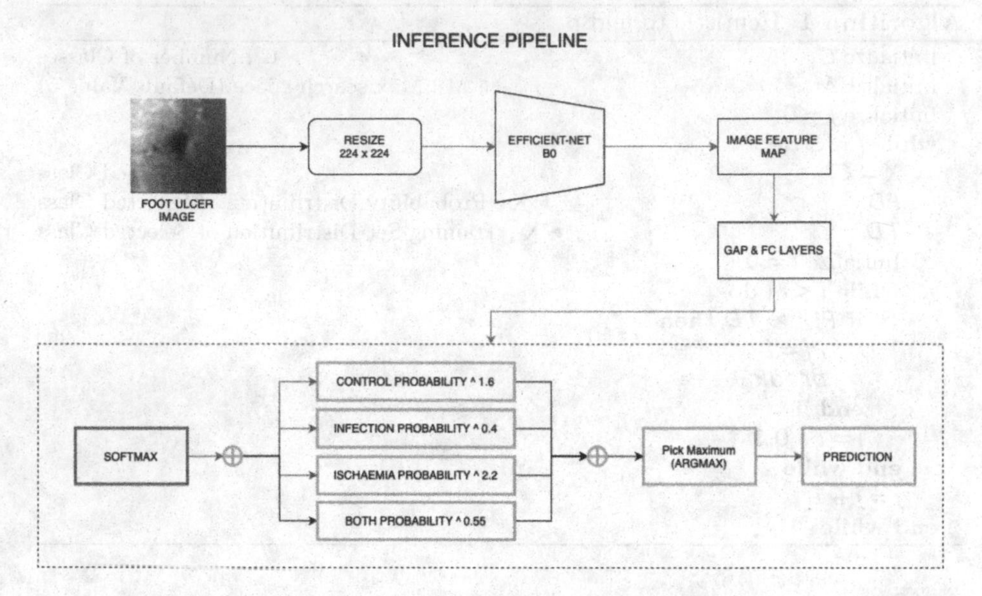

Fig. 5. This figure represents inference pipeline of our pipeline with Bias Adjustable Softmax. We applied Bias Adjustable Softmax with p values of 1.6, 0.4, 2.2, and 0.55 respectively for classes Control, Infection, Ischaemia, and Both.

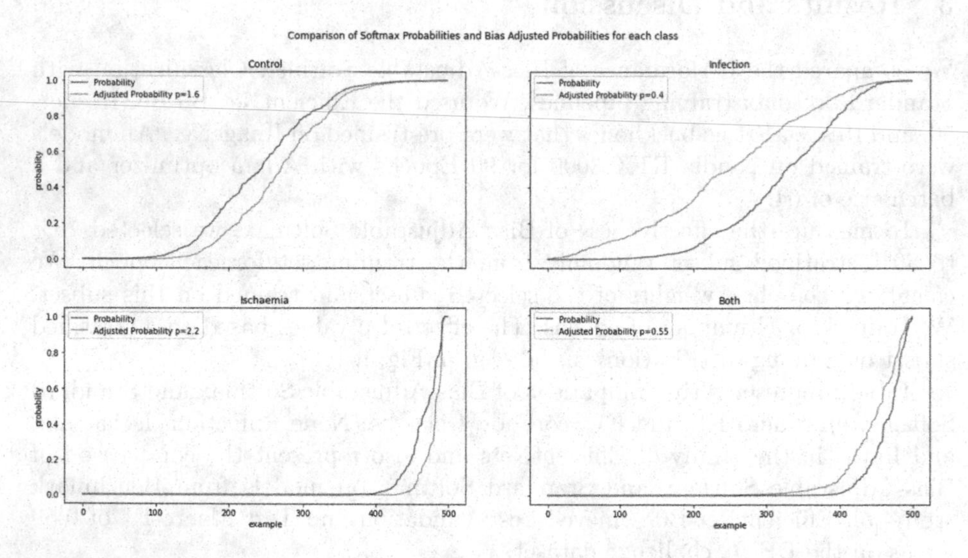

Fig. 6. This figure shows the effect of bias adjustment of the model (EfficientNet-B0) with appropriate p values. There was a significant change in the prediction accuracy for each class. For Control class it decreased from 261 to 218, for Infection class it increased from 177 to 223, for Ischaemia it decreased from 34 to 17, and for Both class it increased from 28 to 42.

Table 2. A comparison of state-of-the-art Image Classification backbones with Bias-Adjustable Softmax and Standard Softmax

Backbone	Adjustable softmax	5-Fold macro F1	Validation macro F1	Validation none F1	Validation infection F1	Validation ischaemia F1	Validation both F1
EfficientNet-B0 (WINNER)	No	0.88	0.5	0.35	0.42	0.46	0.52
	Yes	**0.89**	**0.59**	**0.71**	**0.67**	**0.46**	**0.35**
EfficientNet-B1	No	0.88	0.5	0.42	0.45	0.54	0.54
	Yes	0.88	0.57	0.67	0.66	0.43	0.35
EfficientNet-B2	No	0.89	0.51	0.44	0.49	0.55	0.59
	Yes	0.88	0.55	0.65	0.66	0.54	0.52
EfficientNet-B3	No	0.88	0.49	0.41	0.46	0.54	0.57
	Yes	0.88	0.52	0.66	0.59	0.53	0.54
EfficientNet-B4	No	0.90	0.54	0.66	0.54	0.53	0.64
	Yes	0.90	0.55	0.66	0.51	0.53	0.63
EfficientNet-B5	No	0.87	0.51	0.60	0.64	0.54	0.54
	Yes	0.88	0.53	0.60	0.64	0.54	0.54
EfficientNet-B6	No	0.87	0.53	0.59	0.64	0.55	0.51
	Yes	0.89	0.57	0.68	0.56	0.52	0.65
VGG-16 benchmark	No	0.72	0.51	0.41	0.47	0.52	0.56
	Yes	0.79	0.56	0.60	0.59	0.54	0.64
Inception-V3 benchmark	No	0.85	0.53	0.47	0.58	0.55	0.42
	Yes	0.85	0.54	0.60	0.53	0.49	0.63
Densenet-121 benchmark	No	0.82	0.54	0.51	0.58	0.49	0.51
	Yes	0.88	0.56	0.63	0.55	0.54	0.63
Resnet-101 benchmark	No	0.86	0.51	0.52	0.55	0.40	0.51
	Yes	0.89	0.55	0.60	0.61	0.53	0.67

Figure 7 represents the model performance on validation data by measuring Macro F1-score, and F1-score for 4 classes (None, Infection, Ischaemia, and Both) for each epoch. We observed that almost all models converged to same macro-f1 score in 5-Fold validation. Bias-Adjustable Softmax tends to achieve better results than typical Softmax trained on weighted cross entropy loss. As the image size of DFUC-2021 were 224 × 224 therefore, small models tend to achieve better macro-f1 score than larger models (Recommended image size for EfficientNet-B0 is 224 × 224, and larger models require larger image sizes [11]). As we increase model size, performance tends to decrease because model starts to overfit due to small image size and limited data.

We can use Bias Adjustable Softmax to solve multiple highly imbalanced datasets problems. One of the main applications for Bias Adjustable Softmax is fraud detection where ratio of fraud and normal examples is around 1000:1 [18].

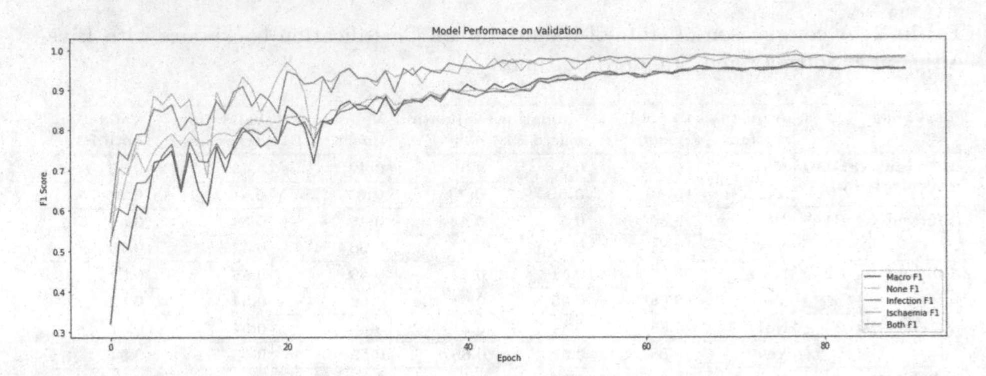

Fig. 7. This figure shows the model learning curve and performance on validation set. Model starts to converge after 50 epochs. F1-score of Ischaemia, and Both classes is higher than None, Infection classes at each epoch despite the less number of examples in these classes. Table 1 shows number of examples and class ratio in dataset.

4 Conclusion

Our experiments show promising results on DFUC2021 dataset. Our Bias-Adjustable Softmax combined with EfficientNet-B0 achieved top-3 rank on the validation leader-board as well as test leaderboard in DFUC2021. Bias adjustable softmax can be used to improve the deep learning models on highly imbalanced datasets.

Acknowledgements. This work was supported by the National Center in Big Data and Cloud Computing (NCBC) and the National University of Computer and Emerging Sciences (NUCES-FAST), Islamabad, Pakistan.

References

1. Wu, S.C., Driver, V.R., Wrobel, J.S., Armstrong, D.G.: Foot ulcers in the diabetic patient, prevention and treatment. Vasc. Health Risk Manag. **3**, 65–76 (2007)
2. Yap, M.H., et al.: Deep learning in diabetic foot ulcers detection: a comprehensive evaluation. Comput. Biol. Med. **135**, 104596 (2021)
3. Goyal, M., Reeves, N.D., Rajbhandari, S., Ahmad, N., Wang, C., Yap, M.H.: Recognition of ischaemia and infection in diabetic foot ulcers: dataset and techniques. Comput. Biol. Med. **117**, 103616 (2020)
4. Rania, N., Douzi, H., Yves, L., Sylvie, T.: Semantic segmentation of diabetic foot ulcer images: dealing with small dataset in DL approaches. In: El Moataz, A., Mammass, D., Mansouri, A., Nouboud, F. (eds.) ICISP 2020. LNCS, vol. 12119, pp. 162–169. Springer, Cham (2020). https://doi.org/10.1007/978-3-030-51935-3_17
5. Yap, M.H., Cassidy, B., Pappachan, J.M., O'Shea, C., Gillespie, D., Reeves, N.D.: Analysis towards classification of infection and ischaemia of diabetic foot ulcers. In: Proceedings of the IEEE EMBS International Conference on Biomedical and Health Informatics (BHI 2021), pp. 1–4 (2021)

6. Cassidy, B., et al.: The DFUC 2020 dataset: analysis towards diabetic foot ulcer detection. touchREVIEWS Endocrinol. **17**, 5–11 (2021). https://www.touchendocrinology.com/diabetes/journal-articles/the-dfuc-2020-dataset-analysis-towards-diabetic-foot-ulcer-detection/1
7. Simonyan, K., Zisserman, A.: Very deep convolutional networks for large-scale image recognition, ArXiv (2014)
8. He, K., Zhang, X., Ren, S., Sun, J.: Deep residual learning for image recognition, ArXiv (2015)
9. Szegedy, C., et al.: Going deeper with convolutions, ArXiv (2014)
10. Huang, G., Liu, Z., van der Maaten, L., Weinberger, K.Q.: Densely connected convolutional networks, ArXiv (2016)
11. Tan, M., Le, Q.V.: EfficientNet: rethinking model scaling for convolutional neural networks, ArXiv (2019)
12. Japkowicz, N., Stephen, S.: The class imbalance problem: a systematic study. Intell. Data Anal. **6**, 429–449 (2002)
13. Hasanin, T., Khoshgoftaar, T.M., Leevy, J.L., Seliya, N.: Examining characteristics of predictive models with imbalanced big data. J. Big Data **6**(1), 1–21 (2019). https://doi.org/10.1186/s40537-019-0231-2
14. Paszke, A., et al.: PyTorch: an imperative style, high-performance deep learning library, ArXiv (2019)
15. Sandler, M., Howard, A., Zhu, M., Zhmoginov, A., Chen, L.-C.: MobileNetV2: inverted residuals and linear bottlenecks, ArXiv (2019)
16. Goyal, P., et al.: Accurate, large minibatch SGD: training ImageNet in 1 hour, ArXiv (2018)
17. Loshchilov, I., Hutter, F.: SGDR: stochastic gradient descent with warm restarts, ArXiv (2017)
18. Phua, C., Lee, V., Smith, K., Gayler, R.: A comprehensive survey of data mining-based fraud detection research, ArXiv (2010)

Efficient Multi-model Vision Transformer Based on Feature Fusion for Classification of DFUC2021 Challenge

Abdul Qayyum[1(✉)], Abdesslam Benzinou[1], Moona Mazher[2], and Fabrice Meriaudeau[3]

[1] ENIB, UMR CNRS 6285 LabSTICC, 29238 Brest, France
qayyum@enib.fr
[2] Department of Computer Engineering and Mathematics, University Rovira i Virgili, Tarragona, Spain
[3] ImViA Laboratory, University of Bourgogne Franche-Comté, Dijon, France

Abstract. Diabetic Foot Ulcers (DFU) is a serious problem and one out of three people with diabetes could be affected by this disease. The rapid rise in the occurrences of DFUs over the last few decades is a major challenge for healthcare systems. DFU with ischemia and infection could be a more serious problem and can cause death. Early detection of DFU and regular monitoring by patients can be useful to overcome the disease. The improvement of patient care and minimization of drawbacks in healthcare systems are also very important. We proposed different pre-trained transformers and fine-tuned them on the DFUC-21 dataset for multi-class classification problems. Before using the Transformer, different pre-trained CNN-based models were fine-tuned on this challenging dataset. When the Transformer was applied to the DFUC-21 dataset, we got favorable results as compared to pre-trained-CNN based models. After several experiments, we have chosen the five best transformers based on pre-trained models and only used two pre-trained transformers in parallel to extract and fuse features from the last layers of the Multi-Model. The proposed model produced Macro-Average F1 0.557 on the validation dataset and 0.569 on a testing dataset and achieved 4th position on the leader board. The proposed model could be useful for the automatic classification of diabetic foot ulcers. The code is publicly available at: https://github.com/RespectKnowledge/DFUC2021.

Keywords: Diabetic Foot Ulcers · Vision Transformer · Multi-model deep learning models · Convolutional neural networks · EfficientNet-B4

1 Introduction

Due to diabetes, 425 million people are globally affected and the expected number would be 629 million by 2045 [1]. Diabetic Foot Ulcers (DFU) is a serious problem and one out of three people with diabetes could be affected by this disease [2]. The rapid rise in the occurrences of DFUs over the last few decades is a major challenge for healthcare

M. H. Yap et al. (Eds.): DFUC 2021, LNCS 13183, pp. 62–75, 2022.
https://doi.org/10.1007/978-3-030-94907-5_5

systems. DFU with ischemia and infection could be a more serious problem and can cause death. A 15–25% chance in a diabetic patient may develop DFU and lower limb amputation could occur if proper foot care and exams are not considered [3]. A diabetic patient requires regular periodic check-ups with doctors, hygienic personal care, and continuous expensive medication to avoid adverse health consequences.

However, in developing countries, the aforementioned caring for a diabetic patient needs a significant amount of time and causes a big financial burden on the patient's family and health services providers and cost of treatment of DFU in developing countries could be equivalent to 5.7 times the annual income [4]. Intelligent automated telemedicine systems would be a better choice to address problems associated with DFU assessment with the construction of Information Communication Technology (ICT) and limited healthcare systems. These telemedicine systems can integrate with current healthcare services to provide more cost-effective, efficient, and quality treatment. These systems can also further improve access to patients from rural and remote backgrounds through the use of the Internet and medical imaging technologies [5]. Early detection of DFU and regular monitoring by patients can be useful to overcome the disease. The improvement of patient care and minimization of drawbacks in healthcare systems are also very important. Recent research on the detection of algorithms can be used to develop a mobile app for use early detection and diagnosis the DFC disease [6, 7]. Nowadays, medical information systems use Computer-Aided Diagnosis (CAD) for detection purposes and in support decision systems as compared to traditional manual data analysis. The main goal is to provide accurate diagnostic support tools for clinicians. Moreover, the current issue is the shortage of specialists for diagnostic tasks in many healthcare institutions.

Traditional methods in machine learning such as Artificial Neural Networks (ANN), Random Forest (RF), and Support Vector Machines (SVM) are widely used for feature extraction and classification tasks [8, 9]. These techniques use relevant properties that are parallel multiprocessing and also use some kernels function for data transformation and require some knowledge about the optimization of various parameters inside these traditional methods. These are providing high-accuracy classification results, especially in the classification task and somehow in the processing of images. However, they require pre-processing steps with handcrafted or manual features extraction and selection, and they also need some domain knowledge about the feature's techniques. The handcrafted features are affected by lighting conditions and skin color depending upon the ethnic group of the patient and hence are not a good choice for reliable automatic detection systems.

Currently, Deep learning (DL) gained wide interest in various tasks such as automatic feature extractions for classification, semantic segmentation, and object detection in medical applications. The shift of attention from conventional paradigms in machine learning to DL is a result of the high accuracy attained through its massive-learning-structures, that allow DL to obtain deeper traits of the data. However, some issues should be addressed in DL when we use it in our design projects. These issues are the data size, the appropriate labeling of the samples, the use of pre-trained structures in the mode of transfer learning, or the design of a proper new learning structure from scratch, and the segmentation and selection of Regions of Interest (ROIs).

In this paper, we have the following contributions:

- The vision-based transformer models have been proposed for DFUC2021 classification. The Multi-Model vision-based models in parallel have been trained and optimized with a weighted cross-entropy function for the classification of multi-class DFUC2021.
- The pair-wise features fusion methods have been used to classify multi-class DFUC2021. The various experiments have been performed with the convolutional neural network as well as transformer-based models using optimizations tricks to tackle the imbalanced class distribution data.

2 Material and Methods

2.1 Dataset

The DFUC2021 challenge organizers provided the diabetic foot ulcer dataset for research purposes. These images are collected during clinical visits of the patients from the Lancashire Teaching Hospitals. Three cameras such as Kodak DX4530, Nikon D3300, and Nikon COOLPIX P100 are used to capture the foot images at a distance of around 30–40 cm with the parallel orientation to the plane of an ulcer. Suitable room lights were used to get the consistent colors in the image. A podiatrist and a consultant physician with specialization with more than 5-year professional experience helped to extract the proper images. The instruction annotation is to identify the location of the ulcer with a bounding box and label each ulcer with ischaemia and/or infection, or none. Similar to DFUC2020 [10], the organizer used the same software, LabelImg [11], to label the images. There are multiple annotations from a podiatrist and a consultant physician with a specialization in the diabetic foot. They average the bounding boxes to form a final bounding box. For diabetic foot pathology classification, the pathology labels were validated with medical records. The two image samples for each class are shown in Fig. 1.

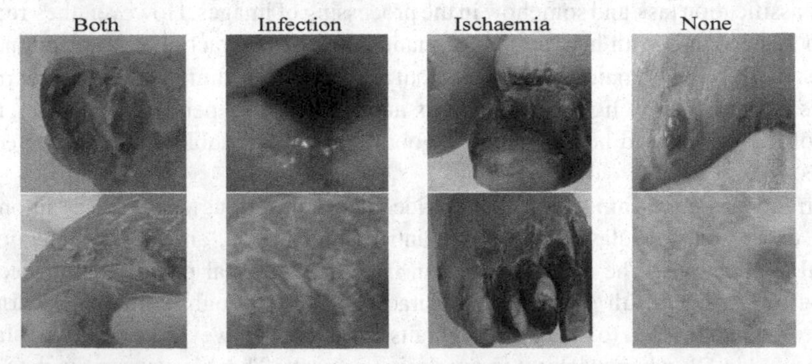

Fig. 1. The data samples were used for testing the proposed model.

The DFUC2021 dataset has been used for training, validation, and testing. Furthermore, the training dataset was split further into 80% for training and 20% for validation.

The total images for DFUC2021 are 15,683, with 5,955 images for the training set (2,555 infections only, 227 ischaemia only, 621 both infection and ischaemia, and 2,552 without ischaemia and infection), 3,994 unlabeled images, and 5,734 images for the testing set. More detail about the dataset can be found in [12].

2.2 Related Work

In computer vision, however, convolutional architectures remain dominant. In the past few years, different researchers and scientists [13–15] have contributed a lot to developing convolutional neural networks for computer vision applications. The most prominent ResNet model has been proposed for computer vision applications and is widely used in image classification, segmentation, and object detection. The mobileNetV1, mobileNetV2 [16] are also used as lightweight models especially for mobile applications and deployments in real-time applications. Recently, The EfficientNet has outperformed in image classification, segmentation and object detection. This architecture used different depth, resolution, and multiscale feature extraction approaches and achieved state-of-the-art results in image recognition and classification. Therefore, the classic ResNet-like architectures are still providing state-of-the-art [17, 18] performance in large-scale image recognition tasks. With the success of natural language processing (NLP), some researchers had tried to combine the CNN-like architectures with self-attention [19, 20] and few replaced entirely CNN-like architectures [21, 22]. These models are theoretically efficient and did not scale effectively on modern hardware accelerators due to the use of specialized attention patterns. In computer vision, the attention is either applied in convolutional networks or used to modify certain components while keeping their overall structure. While the Transformer architecture has become the de-facto standard for natural language processing tasks, its applications to computer vision remain limited. Recently, authors [23] proposed a vision-based transformer without any modification in convolutional neural networks and completely replaced convolutional layers with a transformer. A pure transformer was used directly to sequences of image patches and performed very well on image classification tasks.

Due to the Self-attention-based mechanism, the Transformers [23] has gained popularity in natural language processing (NLP). The Transformer produced better performance in NLP and now gained much popularity in image classification tasks with small modifications. The image is split into patches and provides the sequence of linear imbedding's of these patches as an input to a Transformer. The authors trained the variety of Transformers in a supervised way on the image classification task. The Transformer produced an optimal performance as compared to other CNN-based models. Transformers have lacked some of the inductive biases inherent to CNNs, such as translation variances and locality. Therefore, may not generalize well when trained on an insufficient amount of data. When trained the Vision Transformer (ViT) at sufficient scale and transferred to tasks with fewer data points, it achieved excellent performance. When pre-trained on the public ImageNet-21k dataset or the in-house JFT-300M dataset, ViT approaches or beats state-of-the-art on multiple image recognition benchmarks. In particular, the best model reaches the accuracy of 88.55% on ImageNet, 90.72% on ImageNet-ReaL, 94.55% on CIFAR-100, and 77.63% on the VTAB suite of 19 tasks [23]. The vision-based Transformer trained on multiple datasets (ImageNet, imagenet21k) with different patch sizes

(16,32) and different input resolutions (224,384) is shown in the Fig. 2. and pre-trained weights of all transformers with different patch sizes (16, 32) can be found (https://github.com/rwightman/pytorch-image-models/tree/master/timm).

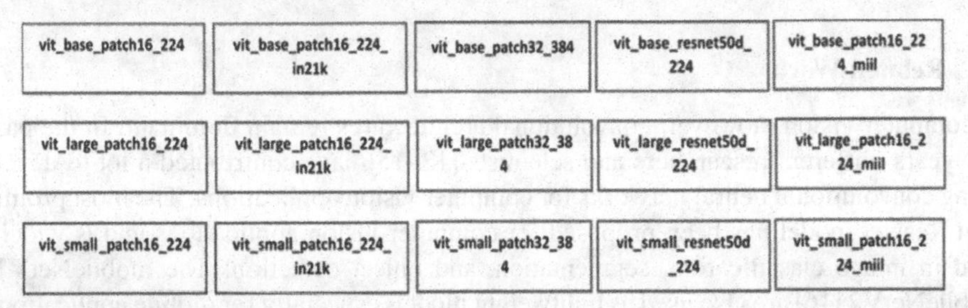

Fig. 2. The vision transformer uses the base, large and small patches with input resolution (224,384), and a hybrid approach (vit-resnet50)

2.3 Proposed Method

2.3.1 Pre-trained Transformers

We have used transformers-based architectures that are originally trained on the ImageNet dataset and fine-tuned these trained transformers by adding some layers at the end of the pre-trained-transformers as shown in Fig. 3. The fully connected (Fc1) with features size (3072 × 768), ReLU activation (ReLU), dropout layer for regularization, and fully connected layer2 (Fc2) with feature size (768 × 4) are used at the end layer of the different pre-trained transformers. The original vision-based transforms are trained on 1000 ImageNet dataset classes. We need to change the last layers to fine-tuned pre-trained transformers according to the number of classes. In our cases, the DFUC21 dataset has four classes.

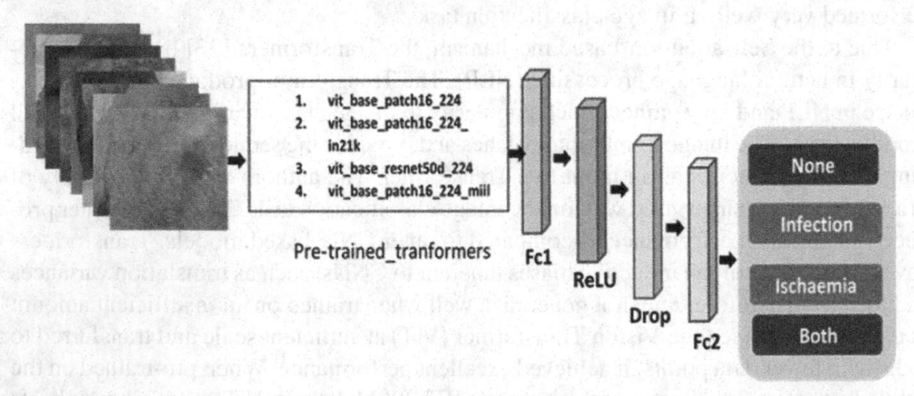

Fig. 3. The different pre-trained transformers were used as fine-tuned by adding a fully connected (Fc1) layer, activation function (ReLU), Dropout layer (Drop), and second fully connected (Fc2) layer for DFUC21 classification.

2.3.2 Proposed Multi Models Vision Transformers with Feature Fusion

In the previous section, we have explained, how can we use pre-trained transformers as fine-tuned for the classification of DFUC21 challenge datasets. We have used different pre-trained transformers and fine-tuned them on the DFUC-21 dataset for multi-class classification problems. Before using the Transformer, the different pre-trained CNN-based models have been fine-tuned on this challenging DFUC21 dataset. When the Transformer was applied to the DFUC-21 dataset, we got favourable results as compared to pre-trained-CNN based models. The different number of transformers with different patch sizes and in hybrid form (combination of vision transformers with ResNet50 backbone) have been fine-tuned. After several experiments, we have chosen the five best transformers based on pre-trained models and did a little trick to improve the performance of the proposed solution. Based on experimental results, the vit_base_patch16_224 architecture produced better performance as compared to other transformers such as vit_base_patch16_224_in21k, vit_base_resnet50d_224 and vit_base_patch16_224_miil [23].

Further, we have proposed the Multi-Model approach with the last feature fusion method. The different transformers have been trained in parallel and features are extracted from the first fully connected layer. Initially, we have used the same model twice in parallel to get features from the last layers and we found the results are not good, then the different models (vit_base_patch16_224, vit_base_patch16_224_miil) in parallel have been trained on the same samples to get different features from each model. The features extracted from the last layer of multiple transformers are concatenated pair-wise and also apply a fully connected layer at the end to concatenate the features of individual transformers and then pass to the classifier for DFUC classification. In pair-wise cross concatenation, first, we concatenate features (F1, F2) and then (F2, F1) and then again concatenate the (F1F2, F2F1). Where F1 features are extracted from model1 and F2 features are extracted from model2. The feature concatenation approach is shown in Fig. 4. The complete proposed block diagram based on a Multi-Model vision transformer is shown in Fig. 5.

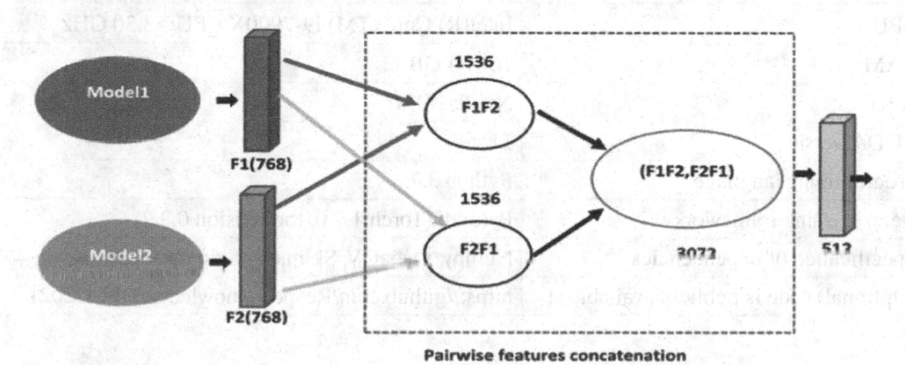

Fig. 4. The pair-wise feature concatenation approach is based on Multi-Model for DFUC21 classification.

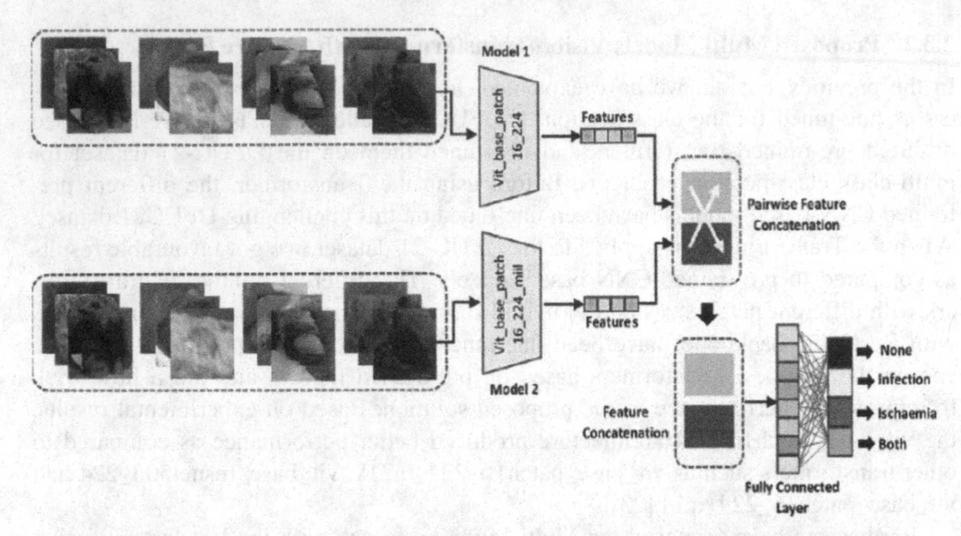

Fig. 5. The proposed Multi-Model architecture for DFUC2021 dataset classification.

2.3.3 Training Parameters and Optimization of the Proposed Model

The PyTorch library is used for model development, training, optimization, testing, and other libraries based on python, are used for pre-processing and analysis of the datasets. The NumPy is used to process the input images. The OpenCV and skimage are used for reading and converting the images into np array from 2D images. The detail of the environment with requirements that are necessary for setting the software and hardware used in training, optimization, and validation of our proposed model is shown in Table 1.

Table 1. Environments and requirements used to train the proposed model

Windows/Ubuntu version	Windows 10
CPU	Intel(R) Core (TM) i9-7900X CPU@3.30 GHz
RAM	16 × 4 GB
GPU	Nvidia V100
CUDA version	11
Programming language	Python 3.7
Deep learning framework	Pytorch (Torch 1.7.0, torchvision 0.2.2)
Specification of dependencies	Numpy, OpenCV, Skimage, Scipy, Pandas
(Optional) code is publicly available at	https://github.com/RespectKnowledge/DFUC2021

The proposed model was trained with different hyperparameters. The learning rate of 0.0001 with Adam optimizer is used for training the proposed model. The weighted cross-entropy function is used as a loss function between the output of the model and the ground-truth sample. The inverse class frequencies for weight balancing have been used to calculate the weighted cross-entropy loss function. The higher-class samples require less weight and fewer class samples need more weight values. The 24 batch-size with 50 epochs has been used with 5 early stopping steps. The best model weights have been saved for prediction in the validation phase. The 224 × 224 input image size was used for training and prediction. The V100 tesla NVidia-GPU machine is used for training and testing the proposed model. The dataset cases have different intensity ranges. The dataset is normalized between 0 and 1 using the max and min intensity normalization method. The detail of the training protocol is shown in Table 2.

Table 2. Training and optimization protocols for the proposed model.

Data augmentation methods	CenterCrop, HorizontalFlip (p = 0.5), VerticalFlip (p = 0.5), RandomBrightnessContrast (p = 0.8), and RandomGamma (p = 0.8)
Initialization of the network	"he" normal initialization
Patch sampling strategy	None
Batch size	24
Patch size	224 × 224
Total epochs	50
Optimizer	Adam
Initial learning rate	0.0001
Learning rate decay schedule	None
Stopping criteria, and optimal model selection criteria	The stopping criterion is reaching the maximum number of epoch (5)
Training time	40 min

3 Performance Metrics

The DFUC2021 dataset has imbalanced distributions of class samples. The most frequently used performance metrics per class are F1-Score, micro average F1-Score, area under the Receiver Operating Characteristics Curve (AUC), macro-average of Precision, Recall, F1-Score, and AUC to reflect the overall performance. The macro-average may be the good choice in imbalanced multi-class settings [12] that would handle cases of strong class imbalance.

The macro-average considers True Positive (TP), False Positives (FP), True Negatives (TN), and False Negatives (FN) for each class *i* (*n* classes), and their respective F1-Scores is shown in Eq. (1). The macro-average F1-Score is determined by averaging the per-class F1-Scores as represented in Eq. (2) [12].

$$\text{Macro-average F1-score: } \frac{1}{N}\sum_{i}^{N} F_1 - Score_i \tag{1}$$

$$\text{Macro-average AUC: } \frac{1}{N}\sum_{i}^{N} Auc_1 \quad (2)$$

Where i = 1,.., N represents the i-th class and N is the total number of classes, in this case, N = 4. The results are computed using macro-average F1-Score. However, other metrics will be discussed, i.e. macro-average recall, macro-average precision, and macro-average AUC.

4 Results and Discussion

The proposed model has been evaluated using the performance metrics provided by the challenge. The average weighted sampling frequency techniques have been also used to handle the imbalance issue. In our experimental results on the validation, the dataset shows better performance with class weighting techniques. The performance metrics reported evaluating the proposed model per class are F1-Score, micro-average F1-Score, and area under the Receiver Operating Characteristics Curve (AUC) and macro-average of Precision, Recall, F1-Score, and AUC. The top results produced by the proposed Multi-Model for each class in the validation and testing phase dataset are shown in Table 3 and Table 4.

Table 3. The performance for each class in the DFUC2021 validation dataset using the proposed Multi-Model vision-based transformer.

Classes	F1 score	Recall	AUC	Precision
None	0.753	0.799	0.842	0.712
Infection	0.661	0.612	0.761	0.720
Ischaemia	0.390	0.533	0.883	0.307
Both	0.423	0.409	0.859	0.439
Weighted-Average	0.671	0.672	–	0.679
Micro-Average	–	–	0.872	–
Macro-Average	**0.557**	0.588	0.837	0.544

Overall, the proposed model produced the highest macro average F1 (0.557) score on the validation dataset and macro average F1 (0.569) on the testing dataset. The proposed model achieved a better macro average AUC score on the validation and test phase dataset. You can see in both Tables 3 and 4, overall Macro average scores for F1, recall, AUC, and precision are increased on the test dataset as compared to the validation dataset. If we check the performance on each class in the dataset, the proposed model did not provide a better solution for the ischaemia class invalidation. However, the model does a better job in the ischaemia class in the test dataset. Also, the performance comparison is not good for the 'Both' class using validation and testing datasets. For 'None' and 'Infection', the proposed model produced better macro average F1 in both validation

Table 4. The performance for each class in the DFUC2021 test dataset using the proposed Multi-Model vision-based transformer.

	F1 score	Recall	AUC	Precision
None	0.746	0.833	0.837	0.675
Infection	0.628	0.548	0.788	0.735
Ischaemia	0.466	0.706	0.919	0.348
Both	0.434	0.352	0.849	0.565
Weighted-Average	0.657	0.663	0.679	0.679
Micro-Average	–	–	0.872	–
Macro-Average	**0.569**	0.610	0.848	0.581

and testing phase datasets. However, if the model is generalized well on the testing dataset that would be helpful towards the first solution for automatic identification of the DFUC2021 classification task.

Table 5. The performance comparison between each class of DFUC2021 test dataset based on proposed and baseline model

Method	Metrics per class F1-score				Micro-average	
	Control	Infection	Ischaemia	Both	F1	AUC
Dens-Net121 [12]	0.72	0.50	0.40	0.36	0.60	0.86
EffNetB0 [12]	0.73	0.56	0.44	0.47	0.63	0.87
Proposed model	0.74	0.62	0.46	0.43	0.65	0.87

Table 6. The overall performance comparison between proposed and baseline model using test DFUC2021 dataset Correction

Method	Macro-average			
	Precision	Recall	F1	AUC
Dens-Net121 [12]	0.56	0.58	0.49	0.85
EffNetB0 [12]	0.57	0.62	**0.55**	0.86
Proposed model	0.58	0.61	**0.56**	0.84

The performance comparison between the proposed and baseline model that was provided by the challenge organizer on the test dataset is shown in Table 5 and Table 6. The proposed model produced a better micro-F1 score in the control and infection class is shown in Table 5. Our proposed model showed a comparatively better macro-F1 score than baseline models shown in Table 6.

Our proposed model's results are comparatively good compared to single transformer-based and convolutional-based neural networks. The ResNet34 [15], DensNet201 [24] and EfficientNet-B4 [25] convolutional neural network has been fine-tuned using some extra layers at the end of the pre-trained models and achieved different F1 scores as shown in Fig. 6. We have also trained and fine-tuned single transformer-based architectures using the DFUC21 dataset and achieved a better macro average F1-score as compared to convolutional neural networks. The comparison between convolutional neural networks with transformers-based models as well as proposed models is shown in Fig. 6. The performance metrics such as Macro-average F1 score, Macro-average Recall, Macro-average AUC and Macro-average precision has been used to compare the proposed and existing deep learning models.

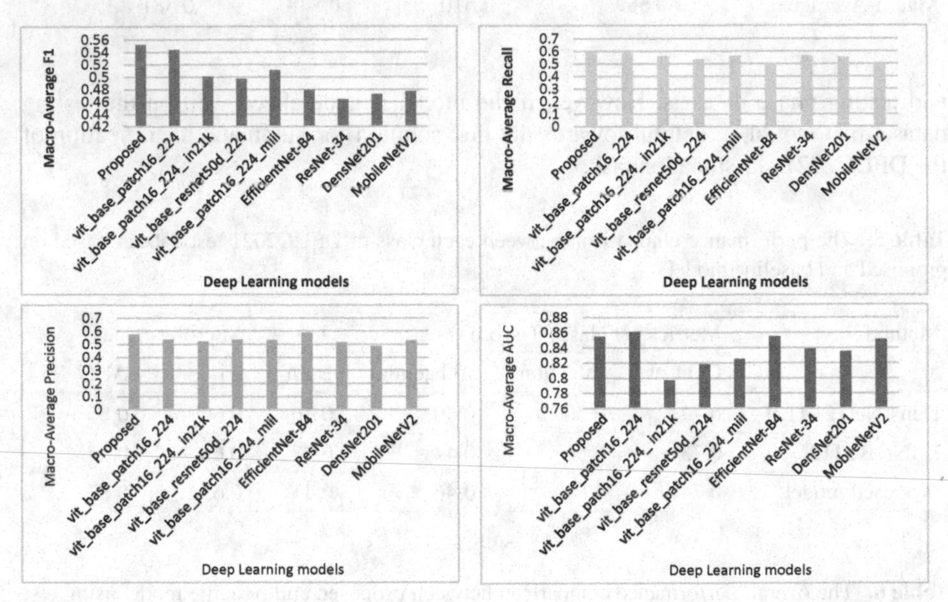

Fig. 6. The comparison between proposed and existing deep learning models for the DFUC2021 validation dataset. The performance has been compared using macro-average F1, Recall, Precision, and AUC for all classes.

The red color at some test sample images is in Fig. 7 shows more activation of gradient-based trained weights of the last layer of the proposed classification model for all classes. We used the weights and gradient of the last layer of our proposed model to visualize the saliency map [26, 27]. This color pattern visualization of the proposed model could better understand the weights activation used for model overfitting interpretation and could be helpful for clinical applications. The notion behind the use of the visualization tool is to achieve better and improved classification scores from the generated image's semantical sections. This also enabled the final output mask for each class to yield higher weights. Figure 7 presents the feature map importance and shows the corresponding right location of features. It would be better to see the feature maps activation of the respective class where the features provide more attention for the good features. The red and green color shows the most important features in a different region of the predicted model. It shows the effectiveness of the proposed model in an explainable way.

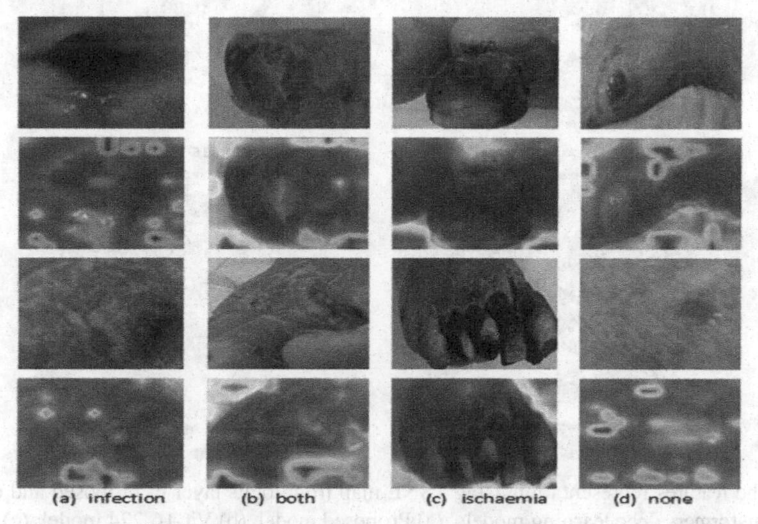

Fig. 7. The feature activation map is based on test samples using the proposed deep learning model. The first row represents the number of samples for class infection, both, ischaemia and none, and the second row represents the activation map. Similarly, the third row represents samples of the same classes and the fourth row represented their activation maps. (Color figure online)

Attention Rollout has been used to visualize the attention maps for two different test cases is represented in the first row and third row, with their respective attention maps are shown in the second and fourth row. The proposed model can make the correct predictions for none class (Fig. 7(d) third and fourth row). Some regions at outlier in the test image did not correspond to the proposed model's actual prediction as shown in Fig. 7(a) and produced a small false positive number of pixels.

The t-Distributed Stochastic Neighbor Embedding (t-SNE) is an unsupervised, non-linear technique primarily used to explore and visualise high-dimensional data. The t-SNE is used to visualize the features from higher dimensional space to lower-dimensional space. The features representation in lower-dimensional space based on proposed and existing deep learning models is shown in Fig. 8. The higher dimension features are extracted from the last layer before the SoftMax layer using proposed and existing transformer-based deep learning models. Figure 8(a) shows the features produced by the proposed model of all four classes are not much overlapped as compared to other transformer-based deep learning models. The proposed model showed strong discriminative power as compared to Vit-16, Vit-16-mili, and Vit-32 models.

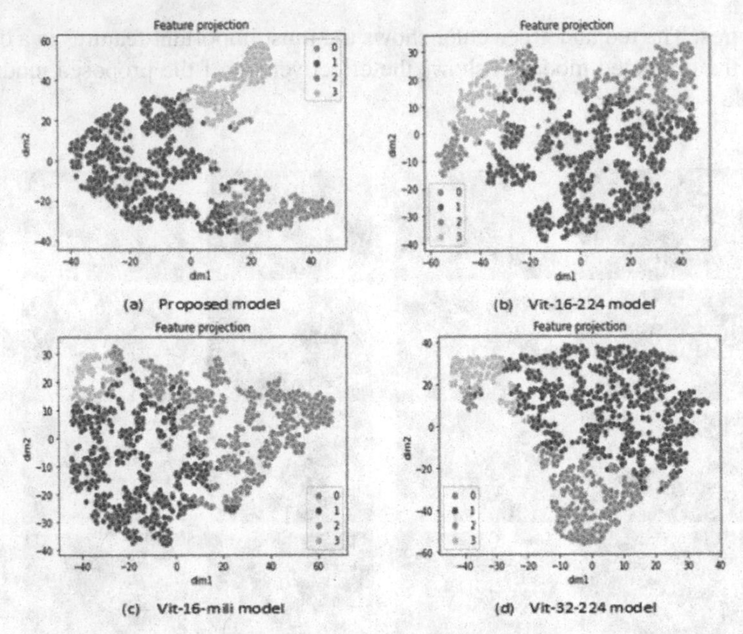

Fig. 8. The features representation using TSNE map from logits layer of proposed and existing vision transformers deep learning models. (a) Proposed model, (b) Vit-16-224 model, (c) Vit-16-mili model and (d) Vit-32-224 model.

5 Conclusion

Few slices have been chosen to see the qualitative information about each class of dataset. In this paper, a vision transformer-based classifier has been proposed with multiple models for the DFUC2021 classification task. The proposed approach produced competitive performance on the test set and achieved a 0.569 macro-average score. The proposed solution would be effective for the automatic classification of DFUC2021 that will be further useful for deployment of the proposed solution in real-time application. In the future, the convolutional neural networks can combine with transformer-based architectures either in feature fusion form or using post ensemble classification scores.

References

1. Cho, N., et al.: IDF diabetes atlas: global estimates of diabetes prevalence for 2017 and projections for 2045. Diabetes Res. Clin. Pract. **138**, 271–281 (2018)
2. Armstrong, D.G., Boulton, A.J., Bus, S.A.: Diabetic foot ulcers and their recurrence. N. Engl. J. Med. **376**(24), 2367–2375 (2017)
3. Aguiree, F., et al.: IDF Diabetes Atlas, 6th edn. International Diabetes Federation, Brussels (2013)
4. Cavanagh, P., Attinger, C., Abbas, Z., Bal, A., Rojas, N., Zhang-Rong, X.: Cost of treating diabetic foot ulcers in five different countries. Diabetes Metab. Res. Rev. **28**, 107–111 (2012)
5. Atzori, L., Iera, A., Morabito, G.: The internet of things: a survey. Comput. Netw. **54**(15), 2787–2805 (2010)

6. Yap, M.H., Chatwin, K.E., Ng, C.-C., Abbott, C.A., Bowling, F.L., Rajbhandari, S., et al.: A new mobile application for standardizing diabetic foot images. J. Diabetes Sci. Technol. **12**(1), 169–173 (2018)
7. Goyal, M., Reeves, N.D., Rajbhandari, S., Yap, M.H.: Robust methods for real-time diabetic foot ulcer detection and localization on mobile devices. IEEE J. Biomed. Health Inform. **23**(4), 1730–1741 (2019)
8. Qayyum, A., Aamir M., Saad, N.M., Mazher. M.: Designing deep CNN models based on sparse coding for aerial imagery: a deep-features reduction approach. Eur. J. Remote Sens. **52**(1), 221–239 (2019)
9. Ren, J. ANN vs. SVM: which one performs better in classification of MCCs in mammogram imaging. Knowl.-Based Syst. **26**, 144–153 (2012)
10. Goyal, M., Yap, M.H., Reeves, N.D., Rajbhandari, S., Spragg, J.: Fully convolutional networks for diabetic foot ulcer segmentation. In: 2017 IEEE International Conference on Systems, Man, and Cybernetics (SMC), pp. 618–623. IEEE (2017)
11. Goyal, M., Hassanpour, S., Yap, M.H.: Region of interest detection in dermoscopic images for natural data-augmentation. arXiv preprint arXiv:1807.10711 (2018)
12. Yap, M.H., Cassidy, B., Pappachan, J.M., O'Shea, C., Gillespie, D., Reeves, N.: Analysis towards classification of infection and ischaemia of diabetic foot ulcers. arXiv preprint arXiv: 2104.03068 (2021)
13. Krizhevsky, A., Sutskever, I., Hinton, G.E.: Imagenet classification with deep convolutional neural networks. In: NIPS (2012)
14. LeCun, Y., et al.: Backpropagation applied to handwritten zip code recognition. Neural Comput. **1**, 541–551 (1989)
15. He, K., Zhang, X., Ren, S., Sun, J.: Deep residual learning for image recognition. In: CVPR (2016)
16. Sandler, M., Howard, A., Zhu, M., Zhmoginov, A., Chen, L.-C.: MobileNetv-2: inverted residuals and linear bottlenecks. In: Proceedings of the IEEE Conference on Computer Vision and *Pattern Recognition*, pp. 4510–4520 (2018)
17. Mahajan, D., et al.: Exploring the limits of weakly supervised pretraining. In: ECCV (2018)
18. Kolesnikov, A., et al.: Big transfer (BiT): general visual representation learning. In: ECCV (2020)
19. Wang, X., Girshick, R., Gupta, A., He, K.: Non-local neural networks. In: CVPR (2018)
20. Carion, N., Massa, F., Synnaeve, G., Usunier, N., Kirillov, A., Zagoruyko, S.: End-to-end object detection with transformers. In: ECCV (2020)
21. Ramachandran, P., Parmar, N., Vaswani, A., Bello, I., Levskaya, A., Shlens, J.: Stand-alone self-attention in vision models. In: NeurIPS (2019)
22. Wang, H., Zhu, Y., Green, B., Adam, H., Yuille, A., Chen, L.C.: Axial-deeplab: stand-alone axial-attention for panoptic segmentation. In: ECCV (2020)
23. Dosovitskiy, A., et al.: An image is worth 16×16 words: transformers for image recognition at scale. arXiv preprint arXiv:2010.11929 (2020)
24. Huang, G., Liu Z., Van Der Maaten, L., Weinberger, K.Q.: Densely connected convolutional networks. In Proceedings of the IEEE Conference on Computer Vision and Pattern Recognition, pp. 4700–4708 (2017)
25. Tan, M., Quoc, L.: EfficientNet: rethinking model scaling for convolutional neural networks. In: International Conference on Machine Learning, pp. 6105–6114. PMLR (2019)
26. Qayyum, A., Razzak, I., Tanveer, M., Kumar, A.: Depth-wise dense neural network for automatic COVID19 infection detection and diagnosis. Ann. Oper. Res. **3**, 1–21 (2021)
27. Almalki, Y.E., et al.: A novel method for COVID-19 diagnosis using artificial intelligence in chest X-ray images. Healthcare **9**(5), 522 (2021)

Classification of Infection and Ischemia in Diabetic Foot Ulcers Using VGG Architectures

Orhun Güley[1,4], Sarthak Pati[1,2,3,4], and Spyridon Bakas[1,2,3](✉)

[1] Center for Biomedical Image Computing and Analytics (CBICA), University of Pennsylvania, Philadelphia, PA, USA

[2] Department of Pathology and Laboratory Medicine, Perelman School of Medicine, University of Pennsylvania, Philadelphia, PA, USA

[3] Department of Radiology, Perelman School of Medicine, University of Pennsylvania, Philadelphia, PA, USA

sbakas@upenn.edu

[4] Department of Informatics, Technical University of Munich, Munich, Germany

Abstract. Diabetic foot ulceration (DFU) is a serious complication of diabetes, and a major challenge for healthcare systems around the world. Further infection and ischemia in DFU can significantly prolong treatment and often result in limb amputation, with more severe cases resulting in terminal illness. Thus, early identification and regular monitoring is necessary to improve care, and reduce the burden on healthcare systems. With that in mind, this study attempts to address the problem of infection and ischemia classification in diabetic food ulcers, in four distinct classes. We have evaluated a series of VGG architectures with different layers, following numerous training strategies, including k-fold cross validation, data pre-processing options, augmentation techniques, and weighted loss calculations. In favor of transparency and reproducibility, we make all the implementations available through the Generally Nuanced Deep Learning Framework (GaNDLF, github.com/CBICA/GaNDLF. Our best model was evaluated during the DFU Challenge 2021, and was ranked 2nd, 5th, and 7th based on the macro-averaged AUC (area under the curve), macro-averaged F1 score, and macro-averaged recall metrics, respectively. Our findings support that current state-of-the-art architectures provide good results for the DFU image classification task, and further experimentation is required to study the effects of pre-processing and augmentation strategies.

Keywords: Diabetic foot · Classification · Deep learning · Convolutional neural network · DFUC2021 · DFU · Ischemia · VGG · GaNDLF

1 Introduction

Diabetic foot ulcers (DFUs) are the most common complications of diabetes mellitus that usually take long time to heal and are among the leading causes

M. H. Yap et al. (Eds.): DFUC 2021, LNCS 13183, pp. 76–89, 2022.
https://doi.org/10.1007/978-3-030-94907-5_6

of hospitalization and morbidity of patients with diabetes [1,2]. According to published estimates, DFU accounts for roughly 20% of hospital admissions of diabetic patients [3]. In addition, DFU leads to substantial emotional, physical, and financial distress that deteriorates the quality of life of patients and their caregivers [4]. If not managed properly, DFU combined with ischemia and infection can cause gangrene, lower limb amputation, and even death [1,2].

For healthcare systems with limited resources, DFU diagnosis can impose a substantial economic burden. As such, the clinical translation of computational methods for the automated assessment of DFU could be beneficial for all stakeholders in the healthcare system, namely, the clinical sites, patients, and caregivers. Such translation could specifically contribute in the early detection of DFU, as well as in the regular monitoring of patients' condition by themselves or by their caregivers. For this purpose, several mobile device applications have been designed and developed for standardization and collection of DFU images, and for promoting self-care of DFU [5–8].

Recent advances in the fields of computer vision and machine learning have had a growing impact on medical imaging, including radiology, histopathology, and dermatology [9–25]. Of special note is the proliferation of several deep learning (DL) methods, which have shown superior performance in numerous computer vision and medical image computing tasks [26–31]. Lately, it is also shown that DL models succeed in classification, detection, and segmentation tasks on DFU images [32–38]. Specifically, previous work from Goyal et al. [33] focuses on the recognition of ischemia and infection on DFU images, but their work tries to solve binary classification problem of ischemia (ischemia vs. non-ischemia) and infection (infection vs. non-infection) separately.

This study aims to solve a multi-class (4-class) classification problem for Diabetic Foot Ulceration (DFU), by leveraging the **G**ener**A**lly **N**uanced **D**eep **L**earning **F**ramework (GaNDLF)[1] [39]. The specific 4 classes considered are: i) *infection*, ii) *ischemia*, iii) *both* infection & ischemia, and iv) *controls* (i.e., neither infection, nor ischemia) (Fig. 1), as defined by the DFU Challenge (DFUC) 2021 [40], conducted in conjunction with the Annual Scientific Meeting of Medical Image Computing and Computer Assisted Interventions (MICCAI) 2021. GaNDLF facilitated our work by providing simple application programming interfaces to rapidly and robustly incorporate techniques such as cross-validation [41], data pre-processing, data augmentation, and weighted loss calculation into our experimental design. Our best model, with which we participated at the DFUC 2021, was compared to the baseline models provided by Yap et al. [40], and was ranked in the 2nd, 5th, and 7th place in the DFUC 2021, based on the macro-averaged AUC (area under the curve), macro-averaged F1 score, and macro-averaged recall metrics, respectively.

[1] https://github.com/CBICA/GaNDLF.

2 Materials and Methods

In this section, we describe the provided dataset in detail, illustrate the examples of infection, ischemia, both infection & ischemia, and control cases from DFU patients with images. Additionally, we explain the methods used in the work, their configuration, the overall VGG architecture, and the various training strategies we followed.

2.1 DFU Dataset

The DFUC2021 dataset describes a multi-institutional collection for analysis of specific pathologies, focusing on infection and ischemia [40]. Specifically, Manchester Metropolitan University and Lancashire Teaching Hospitals established a repository that contains DFU images of infection and ischemia cases for the purpose of supporting research on more advanced methods of pathology detection and recognition of DFU. These DFU images are collected from Lancashire Teaching Hospitals, where photographs were taken from patients during their clinical visits. The three cameras used for capturing the foot images are Kodak DX4530, Nikon D3300 and Nikon COOLPIX P100. The images taken were close-ups of the whole foot from a distance of approximately 30–40 cm with parallel inclination to the ulcer plane. Thereafter, the DFU regions are cropped from the original images and natural data augmentation is performed by preserving the case ids and splitting them to train and test sets.

The complete DFUC2021 dataset comprises of a total of $n = 15,683$ DFU images. The provided ground truth labels, defining the four classes considered by DFUC2021 are i) *infection*, ii) *ischemia*, iii) *both* infection & ischemia, and iv) *controls* (i.e., neither infection, nor ischemia). Representative example cases for each class are shown in Fig. 1. To quantitatively evaluate the performance of algorithms developed for this task, the complete set of $n = 15,863$ images are partitioned into three distinct independent subsets. The training set includes $n = 5,955$ images, provided with their ground truth labels dividing the training set in $n = 2,555$ cases with only infection, $n = 227$ cases with only ischemia, $n = 621$ cases with both infection & ischemia, and $n = 2,552$ control cases. For algorithmic evaluation on unseen data, the DFUC2021 dataset further provides $n = 5,734$ cases for testing, and $n = 3,994$ unlabeled cases. The utilization of unlabelled data for the ischemia and infection classification is left for future work.

2.2 DL Framework

We leveraged the **G**ener**A**lly **N**uanced **D**eep **L**earning **F**ramework (GaNDLF)[2] [39] to conduct all experimentation and training for this study. GaNDLF has been developed in Python using the well-known DL library PyTorch [42]. It enables researchers to target various machine learning (ML) and artificial intelligent (AI)

[2] https://github.com/CBICA/GaNDLF.

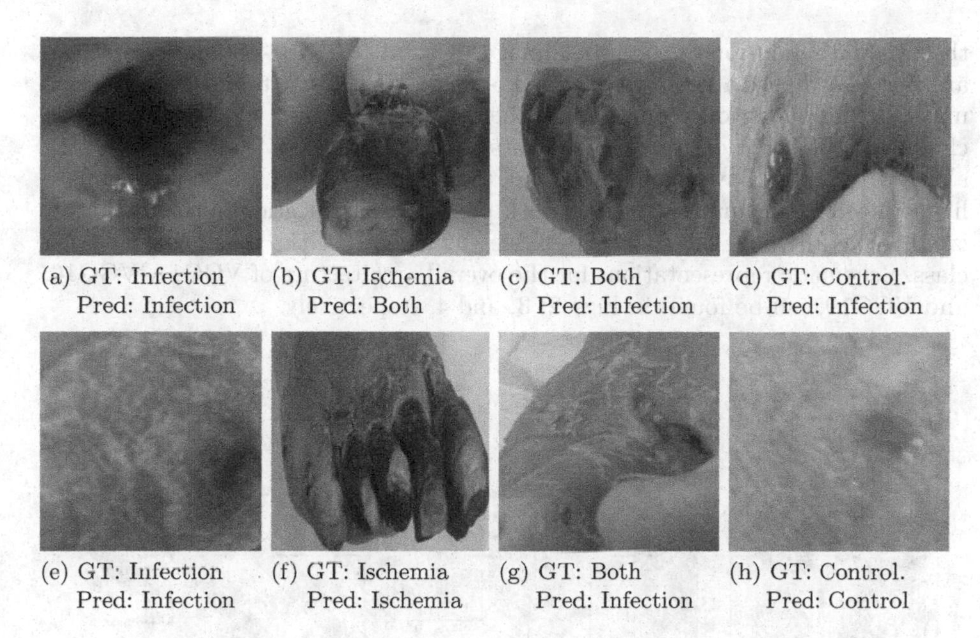

(a) GT: Infection Pred: Infection (b) GT: Ischemia Pred: Both (c) GT: Both Pred: Infection (d) GT: Control. Pred: Infection

(e) GT: Infection Pred: Infection (f) GT: Ischemia Pred: Ischemia (g) GT: Both Pred: Infection (h) GT: Control. Pred: Control

Fig. 1. Examples from each class considered in the DFUC2021 challenge. GT: ground truth, pred: prediction of our best model VGG11 5-fold

workloads (such as segmentation, regression, classification, and synthesis) using different types of imaging modalities (such as RGB, radiographic, and histopathologic imaging), by providing a complete end-to-end solution for training and deploying robust DL models [39]. GaNDLF makes DL accessible for researchers who do not have extensive experience in designing and implementing DL pipelines, while making it straightforward for computational researchers to make their algorithms available for a wider array of applications. It also aims to deploy DL workflows in clinical environments with relative ease. With properties such as an end-to-end application programming interface, ease of training robust and generalizable models with different configurations, robust data pre-processing & augmentation techniques, and nested *k*-fold cross validation, GaNDLF aims to fill gaps of other popular DL libraries. This allows a user to easily design different experiments by simply editing text parameter files without requiring any additional coding, and when combined with the rich metrics library for validating trained models, GaNDLF provides deeper insights into model robustness.

2.3 Architecture

We trained three versions of the VGG architecture [43] for the DFUC2021, namely the VGG11, VGG16, and VGG19 [43]. VGG architectures use very small convolutional filters, and apply spatial padding with the intention of preserving the original resolution of the input image. A total of 5 max-pooling operations are performed over a 2×2 window size, with a stride of 2, to ensure that each max-pooling operation reduces both the image height and width in half. The classifier component of our VGG variants use the ReLU activation function [44], along with a global average pooling [45], two drop-out layers, and a penultimate linear layer. To ensure that the network performs classification, a softmax layer is added as the final layer, which enables us to extract the likelihood for each class. Schematic representations for the overall architecture of VGG11, VGG16, and VGG19, can be found in Figs. 2, 3, and 4, respectively.

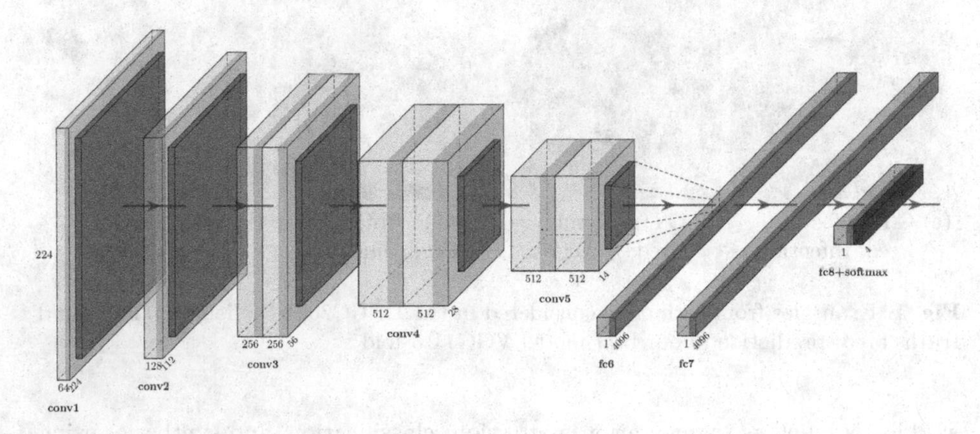

Fig. 2. VGG11 architecture (Figure constructed using PlotNeuralNet [46])

2.4 Training

We used two distinct approaches to train the model. First, we consider a more clinically oriented paradigm, and we split the training data into two equal halves, using one as a retrospective/discovery cohort (i.e., training) and the other as an unseen prospective/replication cohort (i.e., blinded validation). For this partitioning, all subjects of each label/class were proportionally and randomly divided across the 2 halves (retrospective/prospective). This approach was used to yield our baseline results and enabled us to tune the hyperparameters of the model for the task at hand.

Once we obtained these baseline results, we considered a more computationally oriented paradigm, to ensure generalizability of the trained model and prevent overfitting. We specifically employed a k-fold cross validation schema [41], which is a technique widely-used in ML to ensure reporting unbiased performance estimates, and help capture information from an entire dataset by training k different models on k corresponding non-overlapping folds/subsets of the complete training data. Using k-fold cross-validation one can test the model's ability to make accurate predictions on unseen data, detect problems like overfitting or selection bias [47], and provide an understanding on how well the model will generalize to the real distribution of data. For all experiments that used cross-validation, we set the number of folds as $k = 5$. We have performed an equal non-intersecting 5-way split of the training data, ensuring that the model trains on the full training data without overfitting.

Based on the knowledge of the model hyperparameters obtained from our initial experiment using the labelled training data split in equal halves, we proceeded to split the training data into a set of $n = 5$ randomized splits. Each split was used to train the VGG11, VGG16, and VGG19 variants of the network architectures. During the training step, a single patch of 128×128 is extracted from a random location from each image and processed by the model for loss back-propagation. For the forward pass of the model (i.e., the validation/inference phases), enough patches of the specified size are generated from each image to ensure that every pixel is processed at least once by the model, and the final prediction is generated by averaging all the predictions from each patch. To generate the final predictions during the testing phase, we have averaged predictions from every fold. The batch size was chosen as $b = 256$ for VGG11 and VGG16 architectures, and $b = 128$ for VGG19 to ensure maximal utilization of the available hardware resources. For each model architecture, configurations with a standard set of data augmentations and data pre-processing techniques were also evaluated. For data augmentation, bias, blur, noise, and swapping techniques are used with a maximum probability $p = 0.5$. For data pre-processing, a z-scoring normalization mechanism was used. The choice of the loss function stayed the same across these experiments, as the original VGG architecture, which is softmax followed by categorical cross entropy loss, and has shown to work better for multi-class classification problems [48].

To increase the variability of our experiments, we further used weighted cross-entropy loss [49], which ensures that misclassifications of the class with the smallest number of labels generates the largest loss, and could thus be better suited for datasets with imbalanced classes. We used ADAM [50] as the optimizer with an initial learning rate of 0.001. For the half split configurations, the set the total number of epochs to $n = 200$ and set the patience value to $n = 50$. For the 5-fold cross validation configurations, we set the number of epochs to $n = 300$ and patience value to $n = 50$. All of the experimentation was performed on a high performance computing cluster with NVIDIA P100 GPU (which has 11 GB of dedicated video memory), using 32 GB RAM, and 10 CPU threads.

3 Results

In this section, we first analyze the validation performance of the models trained with the different training strategies. We perform our final model selection based on the performance of our partitioned validation set, and submit the inference results from the best model for the challenge evaluation and ranking. Thereafter, we analyze our best models' performance in terms of macro-averaged F1 score, and Macro-AUC, macro-averaged recall and accuracy.

3.1 Training/Validation Dataset Performance

We conducted $n = 12$ different experiments with various training strategies and VGG architectures (Table 1). The models trained without any data augmentation and pre-processing and without using the weighted cross entropy loss outperformed the rest. The overall best model, in terms of validation loss and validation accuracy, was the standard VGG11 architecture trained using 5-fold cross-validation. This model, trained without any data augmentation and pre-processing, was able to reach an average cross entropy loss of 0.24 after averaging over the outputs of all the folds. Table 1 includes a detailed overview of the training results.

Table 1. Validation performance of VGG variations. DA: Data augmentation, DP: Data pre-processing, WCE: Weighted Cross-Entropy. Bold values imply the best performance and the underlined values imply the models that are selected for experimenting in real test set

Model architecture	Training strategy	Validation loss
VGG11	Half-Split	0.49
VGG16	Half-Split	0.51
VGG19	Half-Split	0.59
VGG11	5-fold (DA, DP)	0.52
VGG16	5-fold (DA, DP)	0.63
VGG19	5-fold (DA, DP)	0.72
VGG11	5-fold	**0.24**
VGG16	5-fold	0.30
VGG19	5-fold	0.31
VGG11	5-fold (WCE)	0.27
VGG16	5-fold (WCE)	0.39
VGG19	5-fold (WCE)	0.52

3.2 Testing Dataset Performance

Once we obtained results for all our cross-validated experiments (Table 1), we measured the performance of our best 6 models with the DFUC2021 testing data, as provided by the challenge organizers. The best configuration, in terms of accuracy, macro-averaged F1 score, and Macro-AUC, was the standard VGG11 architecture with 5-fold cross validation. Our best configuration placed in the 5th place in the DFU Challenge 2021, where the participants were ranked according to macro-averaged F1 score. In addition, we analyzed our results according to macro-averaged AUC and macro-averaged recall metrics. Our model ranked in the 2nd and 7th place, respectively (for details, please see the official DFU leaderboard[3]). Considering that the DFUC2021 dataset is an imbalanced set (i.e., the number of samples for each of the classes are not similar), it was surprising to us that standard VGG11 model trained without any pre-processing, data augmentation, and weighted cross-entropy outperformed the other configurations we tested during both validation and testing phases of the challenge. This may require some future meta-analysis to gain a deeper understanding on the exact driving factors. We conducted a visual demonstration of our model's performance on the test set with example samples given by the providers. Samples with ground truth values and predictions can be found in Fig. 1. A detailed illustration of these results can be found at Table 2.

Table 2. Test set performance of the 6 best models, selected according to the average performance of during cross validation training.

Model	Accuracy	Macro-AUC	Macro-Recall	Macro F1-Score
VGG11 5-fold	**0.640**	**0.870**	0.576	**0.561**
VGG16 5-fold	0.617	0.869	0.575	0.543
VGG19 5-fold	0.615	**0.870**	0.572	0.551
VGG11 5-fold WCE	0.624	0.845	0.561	0.541
VGG16 5-fold WCE	0.601	0.858	**0.583**	0.534
VGG19 5-fold WCE	0.595	0.859	0.562	0.521

We also conducted an analysis based on the class-individual F1 scores. The standard VGG11 model performed well based on the ischemia F1 score and control F1 score, achieving ranks of 3rd and 5th, respectively. On the other hand, the performance based on infection F1 score and both F1 score was relatively poor, with rank of 8th for both metrics. Additionally, we compared the VGG19 5-fold model's performance with the other models in the challenge leaderboard and observe that it had the highest 2nd and 5th score based on the results on ischemia F1 score and both F1 score, respectively. Additional details can be found in Table 3.

4 Discussion

In this study, we modified a well-known DL neural network architecture, namely VGG, to classify images containing diabetic foot ulcerations (DFUs), into infection, ischemia, both infection & ischemia, and control. Classification of ischemia and infection of DFU patients is an important task which would help early diagnosis and prevent serious illness in the future. Although 4 classes can be considered as small, the training dataset is not balanced, which makes it harder to learn for the classes with small number of samples. Our results indicate that

[3] https://dfu-2021.grand-challenge.org/evaluation/challenge/leaderboard/.

Table 3. Class-individual F1 scores of the selected models.

Model	Infection F1-Score	Ischemia F1-Score	Both F1-Score	None F1-Score
VGG11 5-fold	**0.547**	0.522	0.440	**0.736**
VGG16 5-fold	0.493	0.531	0.424	0.723
VGG19 5-fold	0.475	**0.548**	0.461	0.719
VGG11 5-fold WCE	0.521	0.488	0.428	0.726
VGG16 5-fold WCE	0.448	0.515	**0.462**	0.712
VGG19 5-fold WCE	0.417	0.502	0.450	0.716

the best approach, in the set of experiments we performed, was the one that did not rely upon weighted loss calculation or any pre-processing methods.

All architectures evaluated in this study were implemented in GaNDLF, which is a high level framework for training robust DL models. For training, we used $n = 12$ different configurations of the general VGG architecture, with different number of weight layers and different training strategies. The number of weight layers were 11, 16 and 19. The two major training strategies we leveraged was 1) splitting the training data into retrospective/training and prospective/-validation datasets, as halves, and 2) following a 5-fold cross validation schema. Effect of data pre-processing and augmentation was explored and quantified in these experiments. It is observed that the 5-fold configuration of the VGG11 without any data augmentation or pre-processing performed the best out all experiments, with an average loss of 0.24 across 5 folds. We hypothesise that VGG16 and VGG19 architectures performed worse than VGG11 because they are simply too large for the given task, and addition of more data is required to properly optimize their weights. The best model in our experimentation was ranked as the 5th place in DFU Challenge 2021, in which the participants were ranked according to macro-averaged F1 score. In addition, our best model was ranked as 2nd and 7th place based on macro-averaged AUC and macro-averaged recall metrics, respectively. Additionally, we further analyzed the class-individual F1 scores and observe that even though our model has performed better on ischemia F1 score and both F1 score, it has performed poorly on infection F1 score and both F1 score metrics.

We believe that there is plenty of room for improvement and further analysis in related future work. Especially, by taking RGB-specific augmentations [51] and pre-processing methods [52] into account, we would expect to significantly improve the model performance. Considering DFU 2021 dataset is imbalanced, exploration of the reasons why weighted cross-entropy loss did not increase the performance would be insightful. Additionally, exploration of custom loss functions, which are specifically designed for optimizing the macro-averaged F1 score could improve the performance [53]. We would also like to see how other popular architectures would perform on this task. Experimentation with architectures like residual networks [54], Efficient-Net [55], Xception-Net [56], InceptionRes-

Net [57] and vision transformers [58] would be insightful. Finally, the use of the unlabelled datasets ($n = 3,994$) to augment the training using weakly supervised training [59] could also help the model performance.

5 Conclusions

DFU can cause critical health problems when it is combined with ischemia and infection. Thus, early diagnosis of potential severe can save lives, and contribute to improvement in the quality of life for all stakeholders in the healthcare system. Automated computational approaches targeting on providing classification suggestions could contribute in early disease detection and the management of DFU patients. Our proposed VGG11 model, trained using a 5-fold cross validation configuration, without any data augmentation or pre-processing, demonstrated superior performance when compared to the other evaluated models, and placed in the 5th place on DFU Challenge 2021, where the rankings were determined by the macro-averaged F1 score. Future work, towards further improving our obtained results, should explore custom loss functions, RGB-specific augmentations (for example, using contrast, brightness, and scale augmentations) along with RGB-specific pre-processing.

Acknowledgments. Research reported in this publication was partly supported by the National Cancer Institute (NCI) of the National Institutes of Health (NIH), under award number NCI:U01CA242871. The content of this publication is solely the responsibility of the authors and does not represent the official views of the NIH.

A Illustrations of Various VGG Variants

A.1 VGG16

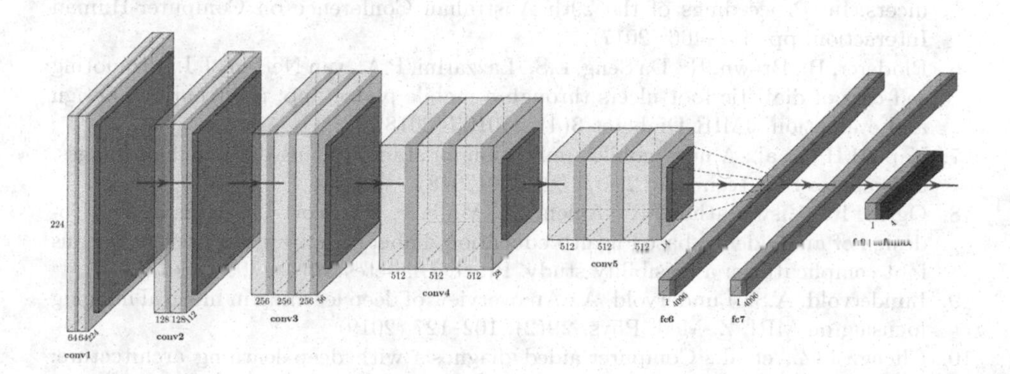

Fig. 3. VGG16 architecture (Figure constructed using PlotNeuralNet [46])

A.2 VGG19

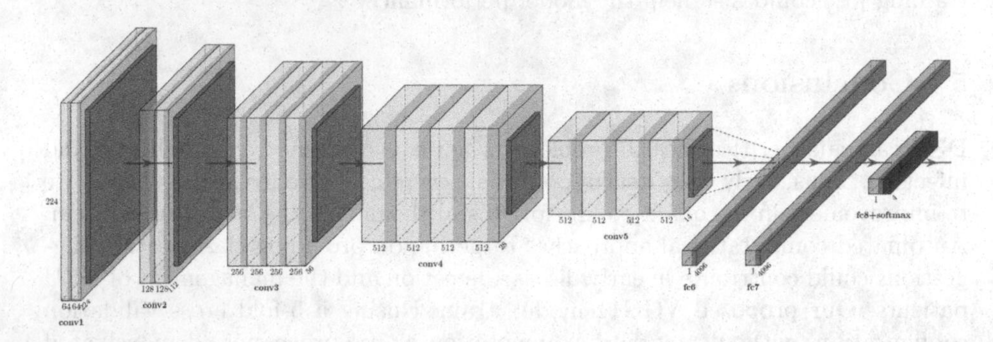

Fig. 4. VGG19 architecture (Figure constructed using PlotNeuralNet [46])

References

1. Yazdanpanah, L., Nasiri, M., Adarvishi, S.: Literature review on the management of diabetic foot ulcer. World J. Diabetes **6**(1), 37 (2015)
2. Shahbazian, H., Yazdanpanah, L., Latifi, S.M.: Risk assessment of patients with diabetes for foot ulcers according to risk classification consensus of international working group on diabetic foot (IWGDF). Pak. J. Med. Sci. **29**(3), 730 (2013)
3. Snyder, R.J., Hanft, J.R.: Diabetic foot ulcers-effects on QOL, costs, and mortality and the role of standard wound care and advanced-care therapies. Ostomy Wound Manage. **55**, 28–38 (2009)
4. Vileikyte, L.: Diabetic foot ulcers: a quality of life issue. Diabetes Metab. Res. Rev. **17**(4), 246–249 (2001)
5. Brown, R., Ploderer, B., Da Seng, L.S., Lazzarini, P., Van Netten, J.: Myfootcare: a mobile self-tracking tool to promote self-care amongst people with diabetic foot ulcers. In: Proceedings of the 29th Australian Conference on Computer-Human Interaction, pp. 462–466 (2017)
6. Ploderer, B., Brown, R., Da Seng, L.S., Lazzarini, P.A., van Netten, J.J.: Promoting self-care of diabetic foot ulcers through a mobile phone app: user-centered design and evaluation. JMIR Diabetes **3**(4), e10105 (2018)
7. Yap, M.H., et al.: A new mobile application for standardizing diabetic foot images. J. Diabetes Sci. Technol. **12**(1), 169–173 (2018)
8. Ogrin, R., Viswanathan, R., Aylen, T., Wallace, F., Scott, J., Kumar, D.: Co-design of an evidence-based health education diabetes foot app to prevent serious foot complications: a feasibility study. Pract. Diabetes **35**(6), 203–209d (2018)
9. Lundervold, A.S., Lundervold, A.: An overview of deep learning in medical imaging focusing on MRI. Z. Med. Phys. **29**(2), 102–127 (2019)
10. Cheng, J.-Z., et al.: Computer-aided diagnosis with deep learning architecture: applications to breast lesions in us images and pulmonary nodules in CT scans. Sci. Rep. **6**(1), 1–13 (2016)
11. Liu, S., et al.: Deep learning in medical ultrasound analysis: a review. Engineering **5**(2), 261–275 (2019)

12. Bakas, S., et al.: Identifying the best machine learning algorithms for brain tumor segmentation, progression assessment, and overall survival prediction in the brats challenge, arXiv preprint arXiv:1811.02629 (2018)
13. Akbari, H., et al.: Histopathology-validated machine learning radiographic biomarker for noninvasive discrimination between true progression and pseudo-progression in glioblastoma. Cancer **126**, 2625–2636 (2020)
14. Akbari, H., et al.: Pattern analysis of dynamic susceptibility contrast-enhanced MR imaging demonstrates peritumoral tissue heterogeneity. Radiology **273**(2), 502–510 (2014)
15. Binder, Z.A., et al.: Epidermal growth factor receptor extracellular domain mutations in glioblastoma present opportunities for clinical imaging and therapeutic development. Cancer Cell **34**(1), 163–177 (2018)
16. Bakas, S., et al.: In vivo detection of egfrviii in glioblastoma via perfusion magnetic resonance imaging signature consistent with deep peritumoral infiltration: the φ-index. Clin. Cancer Res. **23**(16), 4724–4734 (2017)
17. Kurc, T., et al.: Segmentation and classification in digital pathology for glioma research: challenges and deep learning approaches. Front. Neurosci. **14**, 27 (2020)
18. Mang, A., Bakas, S., Subramanian, S., Davatzikos, C., Biros, G.: Integrated biophysical modeling and image analysis: application to neuro-oncology. Annu. Rev. Biomed. Eng. **22**, 309–341, (2020)
19. Bakas, S., et al.: Overall survival prediction in glioblastoma patients using structural magnetic resonance imaging (MRI): advanced radiomic features may compensate for lack of advanced mri modalities. J. Med. Imaging **7**(3), 031505 (2020)
20. Akbari, H. et al.: Survival prediction in glioblastoma patients using multiparametric MRI biomarkers and machine learning methods. ASNR, Chicago, IL (2015)
21. Akbari, H.: et al.: Imaging surrogates of infiltration obtained via multiparametric imaging pattern analysis predict subsequent location of recurrence of glioblastoma. Neurosurgery **78**(4), 572–580 (2016)
22. Akbari, H., Bakas, S., Martinez-Lage, M., et al.: Quantitative radiomics and machine learning to distinguish true progression from pseudoprogression in patients with GBM. In: 56th Annual Meeting of the American Society for Neuroradiology, Vancouver, BC, Canada (2018)
23. Rathore, S., et al.: Radiomic signature of infiltration in peritumoral edema predicts subsequent recurrence in glioblastoma: implications for personalized radiotherapy planning. J. Med. Imaging **5**(2), 021219 (2018)
24. Rathore, S., Bakas, S., Akbari, H., Shukla, G., Rozycki, M., Davatzikos, C.: Deriving stable multi-parametric MRI radiomic signatures in the presence of inter-scanner variations: survival prediction of glioblastoma via imaging pattern analysis and machine learning techniques. In: Medical Imaging 2018: Computer-Aided Diagnosis, vol. 10575, p. 1057509, International Society for Optics and Photonics (2018)
25. Li, H., Galperin-Aizenberg, M., Pryma, D., Simone II, C.B., Fan, Y.: Unsupervised machine learning of radiomic features for predicting treatment response and overall survival of early stage non-small cell lung cancer patients treated with stereotactic body radiation therapy. Radiother. Oncol. **129**(2), 218–226 (2018)
26. Thakur, S., et al.: Brain extraction on MRI scans in presence of diffuse glioma: multi-institutional performance evaluation of deep learning methods and robust modality-agnostic training. Neuroimage **220**, 117081 (2020)

27. Cruz-Vega, I., Hernandez-Contreras, D., Peregrina-Barreto, H., Rangel-Magdaleno, J.d.J., Ramirez-Cortes, J.M.: Deep learning classification for diabetic foot thermograms. Sensors **20**(6), 1762 (2020)
28. Zeng, K., et al.: Segmentation of gliomas in pre-operative and post-operative multimodal magnetic resonance imaging volumes based on a hybrid generative-discriminative framework. In: Crimi, A., Menze, B., Maier, O., Reyes, M., Winzeck, S., Handels, H. (eds.) Brainlesion: Glioma, Multiple Sclerosis, Stroke and Traumatic Brain Injuries, pp. 184–194. Springer, Cham (2016). https://doi.org/10.1007/978-3-319-55524-9_18
29. Sheller, M.J., Reina, G.A., Edwards, B., Martin, J., Bakas, S.: Multi-institutional deep learning modeling without sharing patient data: a feasibility study on brain tumor segmentation. In: Crimi, A., Bakas, S., Kuijf, H., Keyvan, F., Reyes, M., van Walsum, T. (eds.) BrainLes 2018. LNCS, vol. 11383, pp. 92–104. Springer, Cham (2019). https://doi.org/10.1007/978-3-030-11723-8_9
30. Bashyam, V.M., et al.: MRI signatures of brain age and disease over the lifespan based on a deep brain network and 14 468 individuals worldwide. Brain **143**(7), 2312–2324 (2020)
31. Sheller, M.J., et al.: Federated learning in medicine: facilitating multi-institutional collaborations without sharing patient data. Sci. Rep. **10**(1), 1–12 (2020)
32. Goyal, M., Reeves, N.D., Rajbhandari, S., Yap, M.H.: Robust methods for real-time diabetic foot ulcer detection and localization on mobile devices. IEEE J. Biomed. Health Inform. **23**, 1730–1741 (2019)
33. Goyal, M., Reeves, N.D., Rajbhandari, S., Ahmad, N., Wang, C., Yap, M.H.: Recognition of ischaemia and infection in diabetic foot ulcers: dataset and techniques. Comput. Biol. Med. **117**, 103616 (2020)
34. Goyal, M., Yap, M.H., Reeves, N.D., Rajbhandari, S., Spragg, J.: Fully convolutional networks for diabetic foot ulcer segmentation. In: 2017 IEEE International Conference on Systems, Man, and Cybernetics (SMC), pp. 618–623. IEEE (2017)
35. Goyal, M., Reeves, N.D., Davison, A.K., Rajbhandari, S., Spragg, J., Yap, M.H.: DFUNet: convolutional neural networks for diabetic foot ulcer classification. IEEE Trans. Emerg. Top. Comput. Intell. **4**(5), 728–739 (2018)
36. Yap, M.H., et al.: Deep learning in diabetic foot ulcers detection: a comprehensive evaluation. Comput. Biol. Med. **135**, 104596 (2021)
37. Goyal M., Hassanpour, S.: A refined deep learning architecture for diabetic foot ulcers detection, arXiv preprint arXiv:2007.07922 (2020)
38. Alzubaidi, L., Fadhel, M.A., Oleiwi, S.R., Al-Shamma, O., Zhang, J.: DFU QUTNet: diabetic foot ulcer classification using novel deep convolutional neural network. Multimedia Tools and Appl. **79**, 15655–15677 (2019)
39. Pati, S., et al.: GANDLF: a generally nuanced deep learning framework for scalable end-to-end clinical workflows in medical imaging (2021)
40. Yap, M.H., Cassidy, B., Pappachan, J.M., O'Shea, C., Gillespie, D., Reeves, N.: Analysis towards classification of infection and ischaemia of diabetic foot ulcers arXiv preprint arXiv:2104.03068 (2021)
41. Allen, D.M.: The relationship between variable selection and data agumentation and a method for prediction. Technometrics **16**(1), 125–127 (1974)
42. Paszke, A., et al.: PyTorch: an imperative style, high-performance deep learning library. Adv. Neural. Inf. Process. Syst. **32**, 8026–8037 (2019)
43. Simonyan, K., Zisserman, A.: Very deep convolutional networks for large-scale image recognition (2014)
44. Agarap, A.F.: Deep learning using rectified linear units (ReLu), arXiv preprint arXiv:1803.08375 (2018)

45. Lin, M., Chen, Q., Yan, S.: Network in network, arXiv preprint arXiv:1312.4400 (2013)
46. Iqbal, H.: Harisiqbal88/plotneuralnet v1.0.0, December 2018
47. Cawley, G.C., Talbot, N.L.: On over-fitting in model selection and subsequent selection bias in performance evaluation. J. Mach. Learn. Res. **11**, 2079–2107 (2010)
48. Mahajan, D., et al.: Exploring the limits of weakly supervised pretraining. In: Ferrari, V., Hebert, M., Sminchisescu, C., Weiss, Y. (eds.) ECCV 2018. LNCS, vol. 11206, pp. 185–201. Springer, Cham (2018). https://doi.org/10.1007/978-3-030-01216-8_12
49. Fernando, K.R.M., Tsokos, C.P.: Dynamically weighted balanced loss: class imbalanced learning and confidence calibration of deep neural networks. IEEE Trans. Neural Netw. Learn. Syst., 1–12 (2021)
50. Kingma, D.P., Ba, J.: Adam: a method for stochastic optimization, arXiv preprint arXiv:1412.6980 (2014)
51. Buslaev, A., Iglovikov, V.I., Khvedchenya, E., Parinov, A., Druzhinin, M., Kalinin, A.A.: Albumentations: fast and flexible image augmentations. Information **11**(2), 125 (2020)
52. Finlayson, G.D., Schiele, B., Crowley, J.L.: Comprehensive colour image normalization. In: Burkhardt, H., Neumann, B. (eds.) ECCV 1998. LNCS, vol. 1406, pp. 475–490. Springer, Heidelberg (1998). https://doi.org/10.1007/BFb0055685
53. Li, F., Yang, Y.: A loss function analysis for classification methods in text categorization. In: Proceedings of the 20th International Conference on Machine Learning (ICML 2003), pp. 472–479 (2003)
54. He, K., Zhang, X., Ren, S., Sun, J.: Deep residual learning for image recognition. In: Proceedings of the IEEE Conference on Computer Vision and Pattern Recognition, pp. 770–778 (2016)
55. Tan, M., Le, Q.: EfficientNet: rethinking model scaling for convolutional neural networks. In: International Conference on Machine Learning, pp. 6105–6114. PMLR (2019)
56. Chollet, F.: Xception: deep learning with depthwise separable convolutions. In: Proceedings of the IEEE Conference on Computer Vision and Pattern Recognition, pp. 1251–1258 (2017)
57. Szegedy, C., Ioffe, S., Vanhoucke, V., Alemi, A.A.: Inception-v4, inception-ResNet and the impact of residual connections on learning. In: Thirty-First AAAI Conference on Artificial Intelligence (2017)
58. Dosovitskiy, A., et al.: An image is worth 16x16 words: transformers for image recognition at scale. arXiv preprint arXiv:2010.11929 (2020)
59. Wang, D., et al.: Deep-segmentation of plantar pressure images incorporating fully convolutional neural networks. Biocybern. Biomed. Eng. **40**(1), 546–558 (2020)

Diabetic Foot Ulcer Grand Challenge 2021: Evaluation and Summary

Bill Cassidy[1(✉)], Connah Kendrick[1], Neil D. Reeves[2], Joseph M. Pappachan[3], Claire O'Shea[4], David G. Armstrong[5], and Moi Hoon Yap[1]

[1] Department of Computing and Mathematics, Manchester Metropolitan University, Manchester M1 5GD, UK
{B.Cassidy,M.Yap}@mmu.ac.uk

[2] Musculoskeletal Science and Sports Medicine, Manchester Metropolitan University, Manchester M1 5GD, UK

[3] Lancashire Teaching Hospitals NHS Foundation Trust, Preston PR2 9HT, UK

[4] Waikato District Health Board, Hamilton 3240, New Zealand

[5] Southwestern Academic Limb Salvage Alliance (SALSA), Department of Surgery, Keck School of Medicine, University of Southern California, Los Angeles, CA, USA

Abstract. Diabetic foot ulcer classification systems use the presence of wound infection (bacteria present within the wound) and ischaemia (restricted blood supply) as vital clinical indicators for treatment and prediction of wound healing. Studies investigating the use of automated computerised methods of classifying infection and ischaemia within diabetic foot wounds are limited due to a paucity of publicly available datasets and severe data imbalance in those few that exist. The Diabetic Foot Ulcer Challenge 2021 provided participants with a more substantial dataset comprising a total of 15,683 diabetic foot ulcer patches, with 5,955 used for training, 5,734 used for testing and an additional 3,994 unlabelled patches to promote the development of semi-supervised and weakly-supervised deep learning techniques. This paper provides an evaluation of the methods used in the Diabetic Foot Ulcer Challenge 2021, and summarises the results obtained from each network. The best performing network was an ensemble of the results of the top 3 models, with a macro-average F1-score of 0.6307.

1 Introduction

Diabetic foot ulcers (DFU) are one of the most serious complications that can result from diabetes, and often lead to amputation of all or part of a limb due to infection if not met with timely treatment [1,2]. Early detection of DFU, together with accurate screening for infection and ischaemia can help in early treatment and avoidance of further serious complications including amputation. In previous studies, various researchers [3–7] have achieved high accuracy in automated detection of DFUs with machine learning algorithms. A number of widely used clinical DFU classification systems are currently in use, such as

M. H. Yap et al. (Eds.): DFUC 2021, LNCS 13183, pp. 90–105, 2022.
https://doi.org/10.1007/978-3-030-94907-5_7

Wagner [8], University of Texas [9,10], and SINBAD Classification [11], which include information on the site of the DFU, area, depth, presence of neuropathy, ischaemia and infection. We have focused on ischaemia and infection, which are key features of DFU classification systems and important clinical determinants for effective treatment and healing. This focus is consistent with the evolution of threatened limb classification systems, such as the Wound, Ischemia, and foot Infection (WIfI) classification which is used to predict the risk of amputation in patients diagnosed with critical limb ischemia [12,13].

Recognition of infection and ischaemia are key determinate factors that predict the healing progress of DFU and risk of amputation. Ischaemia develops due to lack of arterial inflow to the foot, that results in spontaneous necrosis of the most poorly perfused tissues (gangrene), which may ultimately require amputation of part of the foot or leg. In previous studies, it is estimated that patients with critical limb ischaemia have a three-year limb loss rate of approximately 40% [14]. Patients with an active DFU, particularly those with ischaemia or gangrene, should also be examined for the presence of infection. Approximately, 56% of DFU become infected and 20% of DFU infections lead to amputation of foot or limb [15–17]. In one recent study, 785 million patients with diabetes in the US between 2007 and 2013 suggested that DFU and associated infections constitute a powerful risk factor for emergency department visits and hospital admission [18]. Due to high risks of infection and ischaemia associated with DFU amputation [13], timely and accurate recognition of infection and ischaemia in DFU with cost-effective machine learning methods is an important step towards the development of a complete computerised DFU assessment system.

In current practice, DFU assessment is conducted in foot clinics and hospitals by podiatrists and diabetes physicians. To determine appropriate treatment, a vascular assessment is performed for ischaemia and the wound is assessed for clinical evidence of infection and wound tissue sent for microbiological culture. Van Netten et al. [19] found that clinicians achieved low validity and reliability for remote assessment of DFU in foot images. Hence, it is clear that analysing these conditions from images is extremely difficult even by experienced podiatrists. Patient experiences may be different, however. Swerdlow et al. [20], instituted a "foot selfie" programme and found overall high levels of patient engagement. Limited research exists using computerised methods to automate the monitoring of DFU using foot photographs [21]. This is due to the lack of availability of datasets with clinical labelling for research purposes [22].

Motivated by technological advancements in medical imaging [23–27], where machine learning algorithms performed better than experienced clinicians, Goyal et al. [28] analysed the performance of machine learning algorithms on the recognition of ischaemia and infection on DFUs. They proved that deep learning methods outperformed conventional machine learning methods on a small dataset (1,459 images) and proposed an ensemble Convolutional Neural Network (CNN) approach for ischaemia and infection recognition. Although they achieved high accuracy in ischaemia recognition, there were a number of limitations to their method: 1) the proposed binary classification ensemble CNN method detected one class at a time, which was not capable of detecting co-occurrence of infection and ischaemia; 2) the dataset was small and limits generalisation; 3) the dataset was highly imbalanced, with infection cases significantly outnumbering ischaemic cases; and 4) the recognition rate of infection was 73%, which requires substantial work to improve accuracy. To address these issues, Yap et al. [29] introduced the DFUC2021 datasets, which consist of 4,474 clinically annotated images, together with DFU patches with the label of infection, ischaemia, both infection and ischaemia and none of those conditions (control). Since the release of the DFUC2021 datasets on the 15th April 2021, they have been shared with 51 institutions from 25 countries. Figure 1 illustrates the distribution of researchers using the DFUC2021 datasets by country, showing that the majority of users originate from the United States, China, India and Brazil.

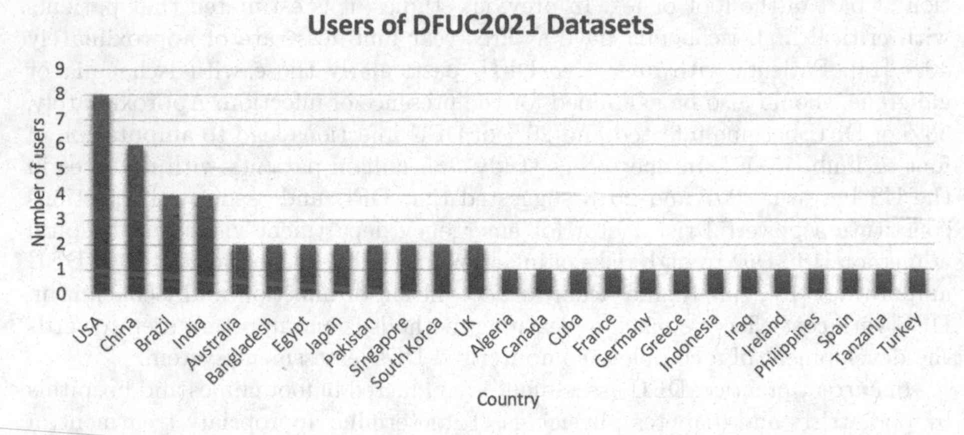

Fig. 1. Distribution of researchers using the DFUC2021 datasets and their country of origin.

2 Methodology

This section summarises the datasets used for DFUC2021, the performance metrics and analysis of the methods proposed by the participants for challenge.

2.1 Datasets and Ground Truth

The previous publicly available dataset created by Goyal et al. [28] consists of 1,459 DFUs: 645 with infection, 24 with ischaemia, 186 with infection and ischaemia, and 604 control DFU (presence of DFU, but without infection or ischaemia). The DFUC2021 dataset is the largest publicly available dataset with DFU pathology labels which consists of 1,703 ulcers with infection, 152 ulcers with ischaemia, 372 ulcers with both conditions, 1703 control DFU and an additional 1,337 unlabelled DFU. The ground truth was produced by two healthcare professionals who specialise in diabetic wounds and ulcers. The instruction for annotation was to label each ulcer with ischaemia, and/or infection, or none.

The patient medical record was used to validate the labels. To increase the number of DFU patches for deep learning algorithms, we used natural augmentation [28] and generated a total of 15,683 DFU patches, which consists of 11,689 labelled patches and 3,994 unlabelled patches. For the labelled DFU patches, the train, validation and test split is: 4,799 patches for the training set, 1,156 patches for the validation set, and 5,734 patches for the testing set. The detailed split for each pathology is presented in Table 1. As shown in Table 1, the number of patches for ischaemia, infection and ischaemia are relatively low when compared to the other classes.

Table 1. DFUC2021 dataset distribution for training (4,799 patches) and validation (1,156 patches) after natural augmentation.

	Infection	Ischaemia	Infection and Ischaemia	None	**Total**
Train	2074	179	483	2063	4799
Validation	481	48	138	489	1156

Figure 2 illustrates DFU patches of four conditions. As these ulcers exhibit variability within a single condition and similarity between different conditions, the DFUC2021 dataset presents a significant challenge for computer vision and machine learning methods in the recognition of infection and ischaemia.

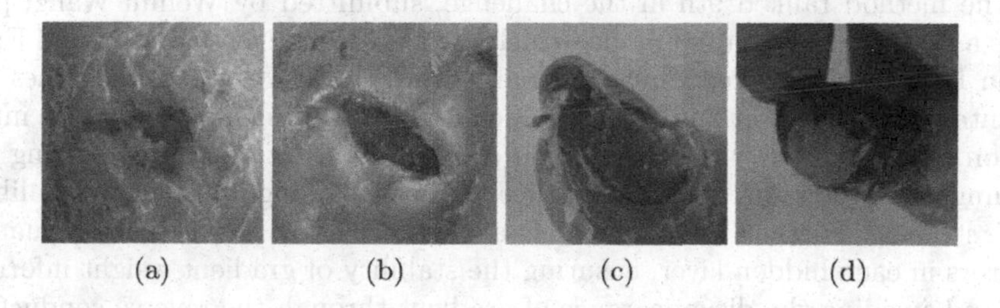

Fig. 2. Illustration of DFU patches from the DFUC2021 dataset. From left to right: (a) control DFU, (b) DFU with infection, (c) DFU with ischaemia and (d) DFU with both infection and ischaemia.

2.2 Performance Metrics

We compared the performance of the deep learning networks on recognition of infection and ischaemia using precision, recall, F1-score and area under the Receiver Operating Characteristics Curve (AUC). For the performance in multi-label classification, due to class imbalance inherent within the DFUC2021 dataset, theperformance will be reported in macro-average. Macro-average is used in imbalanced multi-class settings as it provides equal emphasis on minority classes [30]. To compute macro-average F1-score, first we obtain all the True Positives (TP), False Positives (FP) and False Negatives (FN) for each class i (of n classes), and their respective F1-scores. The Macro-F1 is determined by averaging the per-class F1-scores, $F1_i$:

$$\mathrm{F1}_i = \frac{2 \cdot TP_i}{2 \cdot TP_i + FP_i + FN_i} \tag{1}$$

$$\text{Macro-F1} = \frac{1}{N} \sum_i F1_i \tag{2}$$

where $i = 1, .., N$ represents the i-th class and N is the total number of classes, in this case, $N = 4$. For completeness, we also compared the performance of the algorithms with macro-average AUC.

2.3 Analysis of the Proposed Methods

In this section, we detail the methods used by the top 10 entries for DFUC2021.

The method ranked 10th in the challenge, submitted by Ye Hai, proposes two classifiers. The first classifier is used to detect control cases, while the second classifier is used to detect the other three categories - infection, ischaemia and both infection and ischaemia. The group tested a variety of classification networks, such as ResNet, ViT, DenseNet and SENet, and found that SENet34 provided the best results for both classifiers.

The method ranked 9th in the challenge, submitted by Weilun Wang, proposes a texture classification model which used SE-DenseNet (Squeeze and Excitation Densely Connected Convolutional Network). SE-DenseNet combines the advantages of DenseNet and SENet, which uses multi-dimensional feature information, strengthening the transmission of deep information and enhancing the learning and expression ability of the deep network through a "feature recalibration" strategy. Further, the network is also able to slow down the attenuation of errors in each hidden layer, ensuring the stability of gradient weight information and avoiding the disappearance of gradient through the reverse conduction mechanism of the network itself, improving network performance [31]. This approach does not require a very deep model, so networks such as DenseNet121 and EfficientNetB0, which contain over 100 convolution layers, were not used. To determine the optimum model depth required for this scenario, Wang performed several experiments. First, the 5,955 training samples were split into 10 subsets, followed by 10-fold cross validation. Next, 9-fold samples were randomly up-sampled and used as training samples in each sub-experiment, with the remaining 1-fold samples used for testing.

The method ranked 8th in DFUC2021, submitted by Das et al., proposed a prediction level ensembling. This submission utilised DenseNet121 and EfficientNetB0 models pretrained using ImageNet. The convolutional layers are taken as proposed in the original work, however, the fully connected layers are set as FC(4096), FC(4096), FC(1000) and FC(4), which all use ReLU activations except the final softmax based prediction layer. The configuration of FC layers is motivated by the original VGG16 architecture [32]. The Softmax predictions from both networks are averaged to obtain a prediction level ensembling, providing a final prediction. Figure 3 shows an overview of the network configuration[1].

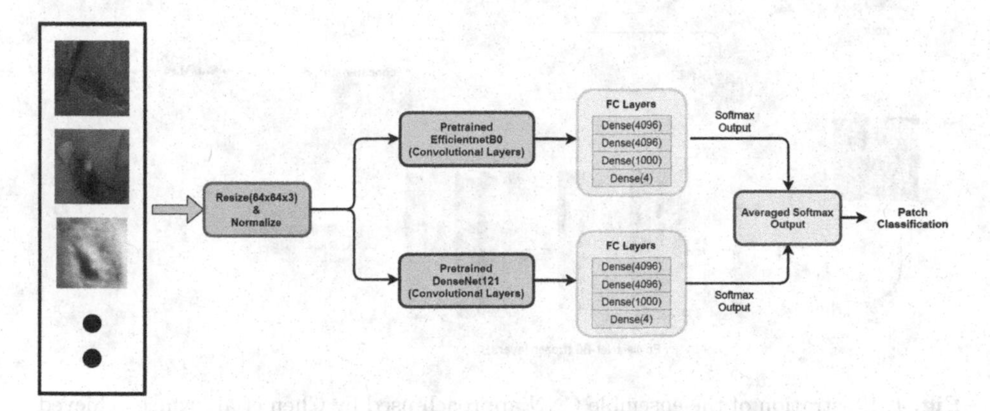

Fig. 3. Illustration of the prediction level ensembling approach used by Das et al., which achieved 8th place in DFUC2021.

The method ranked 7th for DFUC2021, submitted by Chuantao Xie, utlises EfficientNetB0 and DenseNet121, both pretrained using ImageNet. EfficientNetB0 was chosen for its additive feature fusion, while DenseNet121 was chosen for its concatinated feature fusion. The proposed method replaces the layers after the main structure of the CNN, leaving the rest of the network unchanged, including the network structure that proceeds the global average pooling layer which is connected using a parallel structure. One of the branches of the parallel network structure includes convolution, batch normalization, activation function, global average pooling and full connection, followed by the class prediction. Finally, the predicted results of the parallel network structure are concatenated, providing the full connection prediction.

The method which placed 6th for DFUC2021, submitted by Chen et al., utilises an ensemble approach using DenseNet121 and EfficientNet pretrained on ImageNet with a frozen output layer connected to a global average pooling layer. Concatenated integration was implemented with two inputs and one output. The fully connected output layer of the pretrained network was replaced

[1] Reproduced with permission from Sujit Kumar Das, Department of CSE, National Institute of Technology, Silchar 788010, Assam, India.

by a new four-class SoftMax layer. Figure 4 shows an overview of the network configuration[2].

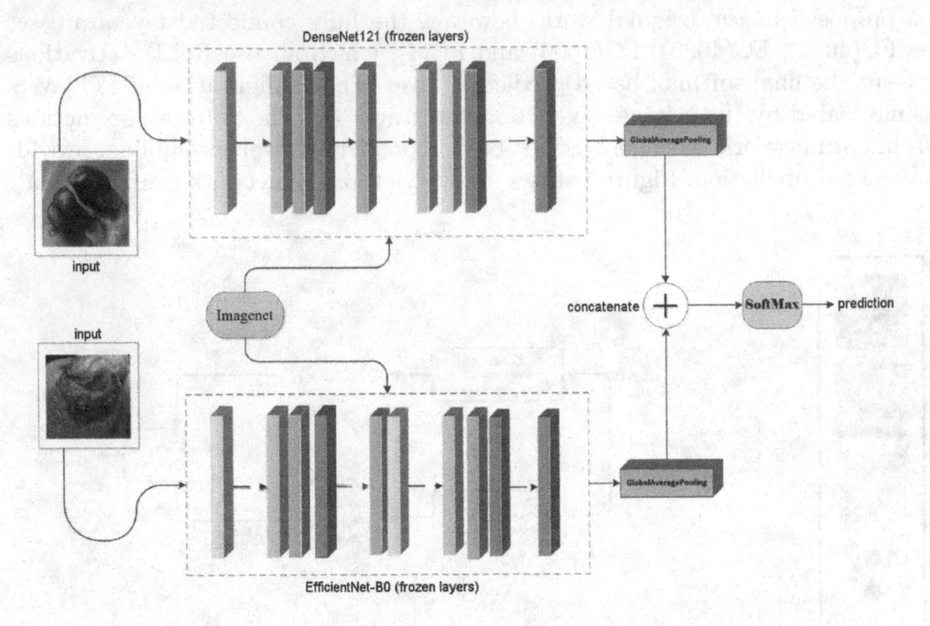

Fig. 4. Illustration of the ensemble CNN approach used by Chen et al., which achieved 6th place in DFUC2021.

The method ranked at 5th place in the challenge, submitted by Güley et al., leveraged the GenerAlly Nuanced Deep Learning Framework (GaNDLF) to achieve multi-class DFU wound classification. GaNDLF enables various machine learning (ML) and artificial intelligent (AI) workloads, including segmentation, regression, classification, and synthesis. This is achieved using a range of imaging modalities, such as RGB, radiographic and histopathologic imaging techniques. Three VGG architectures were trained (VGG11, VGG16 and VGG19) using the DFUC2021 dataset. VGG was selected due to its use of very small convolutions which utilise spacial padding to preserve features from the input image. A total of 5 max-pooling operations were used over a $2 \times N2$ window size, with a stride of 2 to ensure that image dimensions were halved after each max-pooling operation. ReLU activation with global average pooling and two drop-out layers with a penultimate linear layer were used for the classifier. A Softmax layer forms the last layer to provide a final classification.

The method submitted by Qayyum et al., which ranked 4th in the challenge, utilised Vision Transformers (ViTs) to perform DFU classification. ViTs inherently reduce inductive biases, such as translation variances and locality, which

[2] Reproduced with permission from Donghui Lv, Yuqian Chen, School of Communication and Information Engineering, Shanghai University, Shanghai 200444, China.

are present in most other CNN architectures. This solution used pretrained ViTs and fine-tuned them on the DFUC2021 dataset. The features obtained from the last network layer from each ViT were concatenated pairwise, followed by a fully connected layer to concatenate features from individual ViTs before being passed to the final classifier. To address the issue of imbalanced class distribution within the DFUC2021 dataset, average weighted sampling was used, which was shown to improve experimental results.

The method submitted by Ahmed et al., which ranked 3rd in DFUC2021, fine-tuned EfficientNet B0-B6, Resnet-50, and Resnet-101 (pretrained on ImageNet) on the DFUC2021 training set and proposed a custom activation layer using Bias Adjustable Softmax. This Softmax-based activation layer is used to handle class imbalance inherent within the dataset. Their initial experiments used weighted categorical cross entropy but found no significant impact on performance. They found that the use of class weights in the loss function resulted in trained networks showing a bias towards control and infection cases. To address this problem, a novel method was introduced to adjust the skew of probabilities for each class to adjust the bias at inference level.

The method which ranked 2nd place in DFUC2021, submitted by Bloch et al., utilised an ensemble of EfficientNets with a semi-supervised training strategy involving pseudo-labeling for unlabeled images. Their main contribution was the use of Conditional GANs (pix2pixHD) to generate synthetic DFU images to address class imbalance. To achieve this, they created edge masks to indicate regions of interest on the DFUC2021 dataset images.

The winning entry for DFUC2021, submitted by Galdran et al., compared established CNNs (ResNeXt50 and EfficientNet-B3 pretrained on ImageNet) with a Vision Transformer (ViT) and Data-efficient Image Transformers (DeiT) for DFU multi-classification. They also demonstrated how the Sharpness-Aware Minimization (SAM) optimisation algorithm significantly improves the generalisation capability of both traditional CNNs and ViTs in this domain compared to standard Stochastic Gradient Descent (SGD). SAM seeks parameters that lie in neighbourhoods that have uniformly low loss, which results in a min-max optimisation problem on which gradient descent can be performed efficiently, as shown in Algorithm 1. This method was developed to address the problem of heavily overparameterised models where training loss values do not always reflect how well the model generalises. SAM has also been shown to improve robustness against label noise [33]. Their winning entry utilised a linear combination of predictions extracted from BiT-ResNeXt50 (derived from Big Image Transfer) and EfficientNet-B3 models trained on different data folds. This winning submission achieved the highest F1-score (62.16), AUC (88.55) and recall (65.22) measures for DFUC2021.

Algorithm 1. Pseudocode for the SAM algorithm used by the winning entry for DFUC2021, originally proposed by Foret et al. [33].

Input Training set, loss function, batch size, step size, neighborhood size.
Output Model trained with SAM.

1: Initialise weights w_0, $t = 0$
2: **while** *not converged* **do**
3: Sample batch $\beta = \{(x_1, y_1), ...(x_b, y_b)\}$
4: Compute gradient $\nabla_w L_\beta(w)$ of the batch's training loss
5: Compute $\hat{\epsilon}(w)$
6: Compute gradient approximation for the SAM objective: $g = \nabla_w L_\beta(w)|_{w+\hat{\epsilon}(w)}$
7: Update weights: $w_{t+1} = w_t - \eta g$
8: $t = t + 1$
9: return w_t

3 Results and Discussion

For DFUC2021, there were 500 submissions for the validation stage and 28 submissions for the testing stage. Table 2 shows a summary of the top 10 best performing networks submitted to the DFUC2021 test leader board.

3.1 Analysis on the Top-3 Results

In this section, we conduct a statistical analysis of the top-3 results from DFUC2021. Galdran et al. achieved the best F1-score for detection of control cases (0.76), which was an improvement of 0.03 on the baseline (0.73). Ahmed et al. achieved the best F1-score for infection classification (0.68), an improvement of 0.12 on the baseline (0.56). Bloch et al. achieved the best F1-score for ischaemia classification (0.56) which shows an improvement of 0.12 on the baseline (0.44). For F1-score of detection of both infection and ischaemia, Galdran et al. achieved a value of 0.56, resulting in an improvement of 0.09 over the baseline (0.10).

For the micro-average results, Galdran et al. achieved the highest micro-average F1-score (0.68), which is an improvement of 0.05 over the baseline (0.63). They also achieved the highest AUC (0.91), an improvement of 0.04 on the baseline result (0.87).

For the macro-average results, Bloch et al. achieved the highest macro-average precision (0.62) which is an increase of 0.05 over the baseline (0.57). Galdran et al. achieved the best micro-average recall result (0.66), demonstrating an increase of 0.04 over the baseline (0.62). Galdran et al. achieved the highest macro-average F1-score (0.62), which represents an increase of 0.07 over the baseline (0.55). For macro-average AUC, Galdran et al. achieved the best result with (0.89), which is an increase of 0.03 over the baseline result (0.86).

To summarise, the top 3 highest performing entries came from submissions by Galdran et al., Bloch et al. and Ahmed et al. Galdran et al. achieved the highest F1-score for control (0.76), infection and ischaemia (0.57), micro-average F1-score (0.68), micro-average AUC (0.91), and highest macro-average recall (0.66), macro-average F1-score (0.62) and macro-average AUC (0.89). Bloch et al. achieved the highest F1-score for ischaemia classification (0.56), and macro-average precision (0.62). Ahmed et al. achieved the highest F1-score for infection classification (0.68).

The results from the challenge represent a modest increase in performance metrics when compared to the baseline results (mean = 0.064, standard deviation = 0.035, error = 0.011). Possible reasons for this include significant class imbalance inherent within the dataset and the small size of the sample images. F1-scores for infection cases (0.68) and ischaemic cases (0.56) are significantly lower than control cases (0.76), which is a possible further reflection of the class imbalance within the dataset. Additional curation together with additional inter- and intra-rater reliability measures may help to further enhance the datasets. However, dataset curation is a difficult and time-consuming task, and presents additional challenges in the form of label noise and artefacts which could affect the true accuracy of models trained on our data [34–36].

Table 2. Summary of the top 10 performing networks for DFUC2021, compared to the baseline result. AP = Abdul-prediction, AE = Adrian-ensemble, AR = Adrian-results, AM = Ahmed-moded, LE = Louise-ensemble, LID = Louise-ID, ST = shimmer-test, XT = xie-test.

Method	Metrics									
	Per class F1-score				Micro-average		Macro-average			
	Control	Infection	Ischaemia	Both	F1	AUC	Precision	Recall	F1	AUC
Baseline [29]	0.73	0.56	0.44	0.47	0.63	0.87	0.57	0.62	0.55	0.86
AP_vit_bas_GP4EVbn	0.74	0.61	0.43	0.33	0.63	0.87	0.52	0.59	0.53	0.85
AP_vit_mil_UNKBe8A	0.73	0.55	0.52	0.42	0.62	0.87	0.57	0.59	0.56	0.84
AP_vit_multi1_test	0.75	0.63	0.47	0.43	0.66	0.87	0.58	0.61	0.57	0.85
AE_bit_effb3_F2	**0.76**	0.64	0.53	0.56	**0.68**	**0.91**	0.61	0.65	**0.62**	**0.89**
AE_results_final_test	0.74	0.61	0.49	0.49	0.65	0.90	0.58	0.61	0.58	0.88
AE_results_final_test2	**0.76**	0.64	0.51	0.54	0.68	0.90	0.61	0.65	0.61	0.88
AR_final_test4	0.73	0.63	0.52	0.50	0.66	0.90	0.59	0.62	0.60	0.87
AR_final_test5	0.75	0.63	0.51	**0.57**	0.67	0.90	0.61	**0.66**	**0.62**	0.88
AM_v0_89_test	0.72	0.67	0.46	0.54	0.67	0.89	0.60	0.60	0.60	0.86
AM_v0_89_test_1	0.71	**0.68**	0.46	0.53	0.67	0.89	0.60	0.60	0.60	0.86
Arnab	0.73	0.57	0.40	0.45	0.62	0.88	0.53	0.57	0.54	0.85
LE_predictio_aCYsozF	0.75	0.59	**0.56**	0.54	0.65	0.87	**0.62**	0.62	0.61	0.86
LE_Predictio_SSufTEW	0.74	0.60	0.52	0.52	0.65	0.89	0.60	0.62	0.59	0.87
LID47_predictions	0.74	0.58	0.54	0.51	0.65	0.87	0.61	0.61	0.60	0.85
LID48_predictions	0.74	0.59	0.55	0.51	0.65	0.85	0.61	0.62	0.60	0.84
LID49_predictions	0.74	0.59	0.54	0.53	0.65	0.87	0.61	0.63	0.60	0.86
Orhun	0.74	0.55	0.52	0.44	0.62	0.89	0.59	0.58	0.56	0.87
ST_submit1	0.73	0.55	0.48	0.41	0.61	0.87	0.57	0.60	0.54	0.86
ST_submit2	0.74	0.60	0.45	0.43	0.64	0.88	0.56	0.59	0.55	0.86
ST_submit3	0.74	0.57	0.49	0.38	0.63	0.87	0.57	0.59	0.55	0.86
ST_submit4	0.73	0.57	0.46	0.46	0.63	0.87	0.57	0.58	0.56	0.86
Weilunwang	0.70	0.54	0.42	0.47	0.60	0.86	0.54	0.57	0.53	0.82
Yeah	0.72	0.55	0.47	0.35	0.60	0.74	0.53	0.56	0.52	0.70
xie-s_9163	0.72	0.52	0.50	0.46	0.60	0.87	0.57	0.58	0.55	0.86
XT_eff_dense_91_I0yNTTP	0.74	0.56	0.46	0.46	0.63	0.88	0.57	0.60	0.55	0.87

3.2 Ensemble of the Top 10 Performing Models

In this section, we analyse the results of ensembling the top performing models submitted to DFUC2021 to determine if an ensemble approach can provide an increase to performance metrics in multi-classification of DFU. Table 3 shows the results of ensembling the top 10 performing models from DFUC2021.

Table 3. Summary of the results for the top 10 teams and further analysis on the ensembled results. Ensemble Top 2 represents an ensemble of the top 2 teams results, Ensemble Top 3 represents an ensemble of the top 3 teams results, etc.

Method	Metrics									
	Per class F1-score				Micro-average		Macro-average			
	Control	Infection	Ischaemia	Both	F1	AUC	Precision	Recall	F1	AUC
Top-1	0.7574	0.6388	0.5282	0.5619	0.6801	0.9071	0.6140	0.6522	0.6216	0.8855
Top-2	0.7453	0.5917	0.5580	0.5359	0.6532	0.8734	0.6207	0.6246	0.6077	0.8616
Top-3	0.7157	0.6714	0.4574	0.5390	0.6714	0.8935	0.5984	0.5979	0.5959	0.8644
Top-4	0.7466	0.6281	0.4670	0.4347	0.6577	0.8731	0.5814	0.6104	0.5691	0.8488
Top-5	0.7360	0.5468	0.5216	0.4396	0.6199	0.8865	0.5917	0.5759	0.5610	0.8702
Top-6	0.7320	0.5732	0.4621	0.4599	0.6292	0.8725	0.5692	0.5823	0.5568	0.8635
Top-7	0.7407	0.5566	0.4602	0.4558	0.6253	0.8821	0.5705	0.6032	0.5533	0.8698
Top-8	0.7275	0.5701	0.4000	0.4463	0.6222	0.8821	0.5329	0.5692	0.5360	0.8471
Top-9	0.6996	0.5412	0.4237	0.4657	0.5999	0.8622	0.5371	0.5681	0.5326	0.8222
Top-10	0.7192	0.5456	0.4656	0.3532	0.6027	0.7443	0.5300	0.5611	0.5209	0.7020
Ensemble Top 2	0.7455	0.6014	0.5615	0.5301	0.6572	0.9054	0.6187	0.6297	0.6096	0.8866
Ensemble Top 3	0.7491	0.6303	0.5637	0.5799	0.6756	**0.9096**	**0.6352**	**0.6422**	**0.6307**	0.8870
Ensemble Top 4	0.7578	0.6410	0.5513	0.5412	0.6805	0.9102	0.6244	0.6416	0.6228	0.8893
Ensemble Top 5	0.7571	0.6337	0.5653	0.5411	0.6775	0.9112	0.6314	0.6395	0.6243	0.8933
Ensemble Top 6	0.7566	0.6287	0.5437	0.5383	0.6740	0.9104	0.6253	0.6323	0.6168	0.8947
Ensemble Top 7	0.7611	0.6244	0.5417	0.5137	0.6720	0.9099	0.6201	0.6290	0.6102	0.8968
Ensemble Top 8	0.7618	0.6240	0.5486	0.5309	0.6738	0.9106	0.6251	0.6357	0.6163	0.8980
Ensemble Top 9	0.7615	0.6215	0.5629	0.5292	0.6730	0.9093	0.6291	0.6396	0.6188	0.8964
Ensemble Top 10	0.7628	0.6242	0.5603	0.5209	0.6740	0.9082	0.6280	0.6381	0.6171	0.8954

3.3 Visual Comparison of the Top-10 Methods

We conducted further analysis on the top performing methods to determine trends in the data for images that were both easily predicted correctly with high confidence and images where correct classification was difficult. We then visualised those images and identify key features that could have effected the classification result.

Figure 5 highlights 3 images from each class that where correctly classified with high confidence by the top 10 challenge participants. The images show that when the wound is fully visible, the networks are able to determine the difference between classes, even when features from other classes are present, e.g. (a) where the black regions could cause Ischaemia bias. In-contrast Fig. 6 show examples of images that were incorrectly classified by all top 10 challenge participants. These examples highlight the issue with extreme angles and image blur, as seen in images (a), (c), (e) and (f).

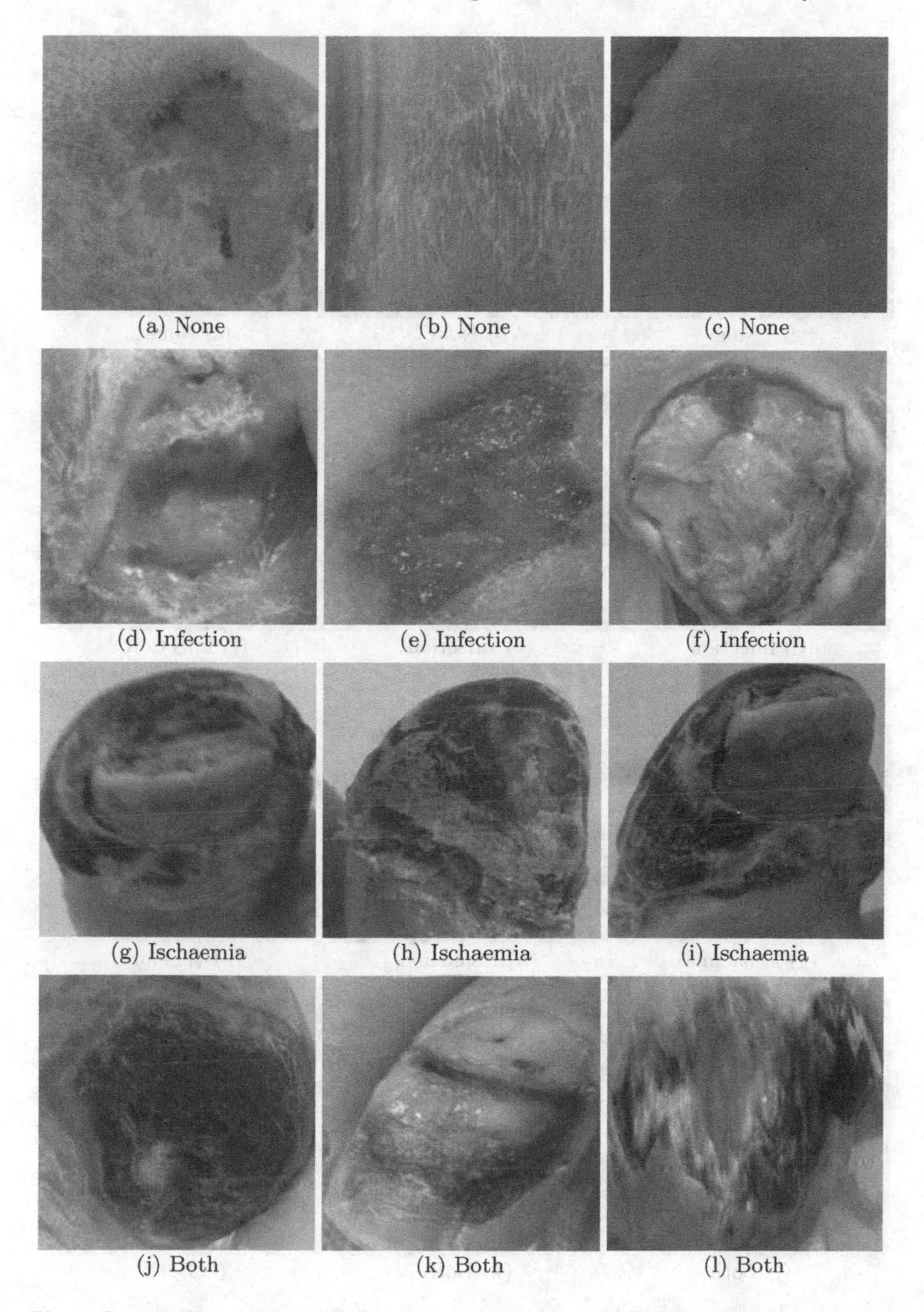

Fig. 5. Images from the testing set which the top 10 networks all predicted correctly.

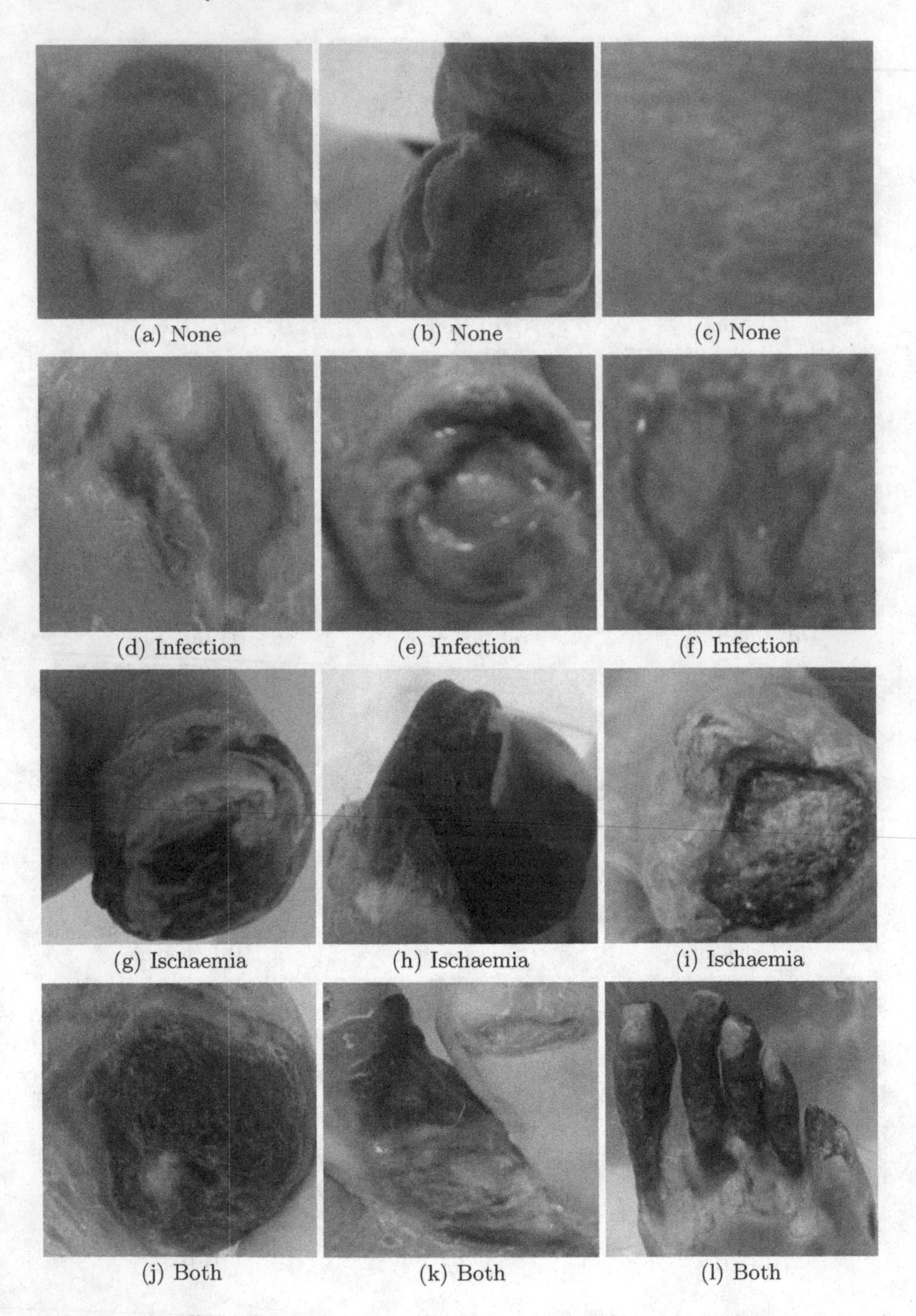

(a) None (b) None (c) None

(d) Infection (e) Infection (f) Infection

(g) Ischaemia (h) Ischaemia (i) Ischaemia

(j) Both (k) Both (l) Both

Fig. 6. Images from the testing set which the top 10 networks all predicted incorrectly.

One particularly notable result can be seen when comparing the correctly classified image (j) in Fig. 5 with the incorrectly classified image (j) in Fig. 6. The image in Fig. 6 is the result of subtle natural augmentation and has resulted in an incorrect classification.

4 Conclusion

In this study, we introduce the largest international DFU pathology dataset, and propose a weakly supervised framework for DFU pathology classification of infection and ischaemia. This is the first dataset of its kind to be made available to the research community together with implementation of CNNs for multi-class classification of infection, ischaemia, and co-occurrences of infection and ischaemia. These advancements will help to support early identification of DFU complications to guide treatment and help prevent further complications including limb amputation.

Although the majority of the deep learning methods reported in this paper show promising results in recognising infection and ischaemia, there are still significant challenges in designing methods to detect the co-occurrences of both conditions. Future work will investigate more advanced techniques such as generative adversarial networks and unsupervised learning to improve network performance.

This work forms an important contribution to our ongoing research into developing a fully automated DFU diagnosis and monitoring framework which can be used by patients and their carers in home settings, to help reduce strains on healthcare services around the world. This work will build on our existing framework [37,38] in delivering an easy-to-use system capable of advanced forms of diabetic foot analysis, which will include longitudinal monitoring as a means of assessing wound healing progress.

Acknowledgment. We gratefully acknowledge the support of NVIDIA Corporation who provided access to GPU resources for the DFUC2021 Challenge and an NVIDIA Geforce RTX 3090 GPU card as the prize for the winning team.

References

1. Armstrong, D.G., Boulton, A.J.M., Bus, S.A.: Diabetic foot ulcers and their recurrence. N. Engl. J. Med. **376**(24), 2367–2375 (2017)
2. Boulton, A.J.M., et al.: Diagnosis and management of diabetic foot complications (2019)
3. Wang, C., et al.: A unified framework for automatic wound segmentation and analysis with deep convolutional neural networks. In: Engineering in Medicine and Biology Society (EMBC), 2015 37th Annual International Conference of the IEEE, pp. 2415–2418. IEEE (2015)
4. Goyal, M., Yap, M.H., Reeves, N.D., Rajbhandari, S., Spragg, J.: Fully convolutional networks for diabetic foot ulcer segmentation. In: 2017 IEEE International Conference on Systems, Man, and Cybernetics (SMC), pp. 618–623, October 2017

5. Goyal, M., Reeves, N.D., Davison, A.K., Rajbhandari, S., Spragg, J., Yap, M.H.: DFUNet: convolutional neural networks for diabetic foot ulcer classification. IEEE Trans. Emerg. Top. Comput. Intell. **4**, 1–12 (2018)
6. Goyal, M., Reeves, N.D., Rajbhandari, S., Yap, M.H.: Robust methods for real-time diabetic foot ulcer detection and localization on mobile devices. IEEE J. Biomed. Health Inform. **23**(4), 1730–1741 (2019)
7. Yap, M.H., et al.: Deep learning in diabetic foot ulcers detection: a comprehensive evaluation. Comput. Biol. Med. **135**, 104596 (2021)
8. Wagner, F.W.: The diabetic foot. Orthopedics **10**(1), 163–172 (1987)
9. Lavery, L.A., Armstrong, D.G., Harkless, L.B.: Classification of diabetic foot wounds. J. Foot Ankle Surg. **35**(6), 528–531 (1996)
10. Armstrong, D.G., Lavery, L.A., Harkless, L.B.: Validation of a diabetic wound classification system: the contribution of depth, infection, and ischemia to risk of amputation. Diabetes Care **21**(5), 855–859 (1998)
11. Ince, P., et al.: Use of the SINBAD classification system and score in comparing outcome of foot ulcer management on three continents. Diabetes Care **31**(5), 964–967 (2008)
12. Armstrong, D.G., Mills, J.L.: Juggling risk to reduce amputations: the three-ring circus of infection, ischemia and tissue loss-dominant conditions. Wound Med. **1**, 13–14 (2013)
13. Mills, J.L., Sr., et al.: The society for vascular surgery lower extremity threatened limb classification system: risk stratification based on wound, ischemia, and foot infection (wifi). J. Vasc. Surg. **59**(1), 220–234 (2014)
14. Albers, M., Fratezi, A.C., De Luccia, N.: Assessment of quality of life of patients with severe ischemia as a result of infrainguinal arterial occlusive disease. J. Vasc. Surg. **16**(1), 54–59 (1992)
15. Prompers, L., et al.: High prevalence of ischaemia, infection and serious comorbidity in patients with diabetic foot disease in Europe baseline results from the Eurodiale study. Diabetologia **50**(1), 18–25 (2007)
16. Lipsky, B.A., et al.: 2012 infectious diseases society of America clinical practice guideline for the diagnosis and treatment of diabetic foot infections. Clin. Infect. Dis. **54**(12), e132–e173 (2012)
17. Lavery, L.A., Armstrong, D.G., Wunderlich, R.P., Tredwell, J., Boulton, A.J.M.: Diabetic foot syndrome: evaluating the prevalence and incidence of foot pathology in Mexican Americans and non-hispanic whites from a diabetes disease management cohort. Diabetes Care **26**(5), 1435–1438 (2003)
18. Skrepnek, G.H., Mills, J.L., Lavery, L.A., Armstrong, D.G.: Health care service and outcomes among an estimated 6.7 million ambulatory care diabetic foot cases in the US. Diabetes Care **40**(7), 936–942 (2017)
19. van Netten, J.J., Clark, D., Lazzarini, P.A., Janda, M., Reed, L.F.: The validity and reliability of remote diabetic foot ulcer assessment using mobile phone images. Sci. Rep. **7**(1), 9480 (2017)
20. Swerdlow, M., Shin, L., D'Huyvetter, K., Mack, W.J., Armstrong, D.G.: Initial clinical experience with a simple, home system for early detection and monitoring of diabetic foot ulcers: the foot selfie. J. Diabetes Sci. Technol. (2021)
21. Yap, M.H., et al.: A new mobile application for standardizing diabetic foot images. J. Diabetes Sci. Technol. **12**(1), 169–173 (2018)
22. Cassidy, B., et al.: The DFUC 2020 dataset: analysis towards diabetic foot ulcer detection. touchREVIEWS Endocrinol. **17**, 5–11 (2021)
23. Esteva, A., et al.: Dermatologist-level classification of skin cancer with deep neural networks. Nature **542**, 01 (2017)

24. Brinker, T.J., et al.: Deep neural networks are superior to dermatologists in melanoma image classification. Eur. J. Cancer **119**, 11–17 (2019)
25. Fujisawa, Y., et al.: Deep-learning-based, computer-aided classifier developed with a small dataset of clinical images surpasses board-certified dermatologists in skin tumour diagnosis. Br. J. Dermatol. **180**(2), 373–381 (2019)
26. Pham, T.C., et al.: Improving binary skin cancer classification based on best model selection method combined with optimizing full connected layers of deep CNN. In: 2020 International Conference on Multimedia Analysis and Pattern Recognition (MAPR), pp. 1–6 (2020)
27. Jinnai, S., Yamazaki, N., Hirano, Y., Sugawara, Y., Ohe, Y., Hamamoto, R.: The development of a skin cancer classification system for pigmented skin lesions using deep learning. Biomolecules **10**(8), 1123 (2020)
28. Goyal, M., Reeves, N.D., Rajbhandari, S., Ahmad, N., Wang, C., Yap, M.H.: Recognition of ischaemia and infection in diabetic foot ulcers: dataset and techniques. Comput. Biol. Med. **117**, 103616 (2020)
29. Yap, M.H., Cassidy, B., Pappachan, J.M., O'Shea, C., Gillespie, D., Reeves, N.: Analysis towards classification of infection and ischaemia of diabetic foot ulcers. arXiv preprint arXiv:2104.03068 (2021)
30. Forman, G., Scholz, M.: Apples-to-apples in cross-validation studies: pitfalls in classifier performance measurement. ACM SIGKDD Explor. Newsl. **12**(1), 49–57 (2010)
31. Jingyi, Q., Zhao, T., Ye, M., Li, J., Liu, C.: Flight delay prediction using deep convolutional neural network based on fusion of meteorological data. Neural Process. Lett. **52**, 10 (2020)
32. Simonyan, K., Zisserman, A.: Very deep convolutional networks for large-scale image recognition. CoRR, abs/1409.1556 (2015)
33. Foret, P., Kleiner, A., Mobahi, H., Neyshabur, B.: Sharpness-aware minimization for efficiently improving generalization. arXiv preprint arXiv:2010.01412 (2021)
34. Wen, D., et al.: Characteristics of publicly available skin cancer image datasets: a systematic review. Lancet Digit. Health **4** (2021)
35. Cassidy, B., Kendrick, C., Brodzicki, A., Jaworek-Korjakowska, J., Yap, M.H.: Analysis of the ISIC image datasets: usage, benchmarks and recommendations. Med. Image Anal. **75** (2021)
36. Daneshjou, R., et al.: Checklist for evaluation of image-based artificial intelligence reports in dermatology: CLEAR derm consensus guidelines from the international skin imaging collaboration artificial intelligence working group. JAMA Dermatol. (2021)
37. Reeves, N.D., Cassidy, B., Abbott, C.A., Yap, M.H.: Chapter 7 - novel technologies for detection and prevention of diabetic foot ulcers. In: Gefen, A. (ed.) The Science, Etiology and Mechanobiology of Diabetes and its Complications, pp. 107–122. Academic Press (2021)
38. Cassidy, B., et al.: A cloud-based deep learning framework for remote detection of diabetic foot ulcers. arXiv preprint arXiv:2004.11853 (2021)

Post Challenge Paper

Deep Subspace Analysing for Semi-supervised Multi-label Classification of Diabetic Foot Ulcer

Azadeh Alavi[1,2(✉)] and Hossein Akhoundi[3]

[1] AI Discipline, School of Computing Technologies, RMIT University, Melbourne, VIC 3001, Australia
azadeh.alavi@rmit.edu.au
[2] Bioinformatics Lab, Baker Heart and Diabetes Institute, Melbourne, VIC 3002, Australia
[3] CISCO Systems Australia, 4/101 Collins Street, Melbourne, VIC 3000, Australia

Abstract. Diabetes is a global raising pandemic. Diabetic patients are at risk of developing foot ulcer that usually leads to limb amputation. In order to develop a self monitoring mobile application, in this work, we propose a novel deep subspace analysis pipeline for semi-supervised diabetic foot ulcer mulit-label classification. To avoid risk of over-fitting, the proposed pipeline dose not include any data augmentation. Whereas, after extracting deep features, and in order to make the representation shift invariant, we employ variety of data augmentation methods on each image and generate image-sets, which are then mapped into linear subspaces. Moreover, the proposed pipeline reduces the cost of retraining when more new unlabelled data become available. Thus, the first stage of the pipeline employs the concept of transfer learning for feature extraction purpose through modifying and retraining a deep convolutional network architect known as Xception. Then, the output of a mid-layer is extracted to generate an image set representer of any given image with help of data augmentation methods. At this stage, each image is transferred to a linear subspace which is a point on a Grassmann Manifold topological space. Hence, to perform analyses, the geometry of such manifold must be considered. As such, inspired by the "relational divergence" method, each labelled image is represented as a relational vector. That is through calculating their geodesic distance on Grassmann manifold to number of unlabelled patches. Finally, Random Forest is trained for multi-label classification of diabetic foot ulcer images. The method is then evaluated on the blind test set provided by DFU2021 competition, and the result considerable improvement compared to using classical transfer learning with data augmentation.

Keywords: Deep learning · Semi-supervised · Medical images · DFUC2021 · Diabetic foot ulcer

---

Professor David Ascher's lab: david.ascher@baker.edu.au.

M. H. Yap et al. (Eds.): DFUC 2021, LNCS 13183, pp. 109–120, 2022.
https://doi.org/10.1007/978-3-030-94907-5_8

1 Introduction

Diabetes is a raising universal problem that affects 425 million people which is expected to rise to 629 million people by 2045 [7,18]. One in three diabetic patients are likely to develop Diabetic Foot Ulcers (DFU) which is a serious complication of diabetes, and can lead to limb amputation, or even death, owing to complications, such as infection and ischaemia [3]. However, as diabetic patients normally loose sensation in their foot, it is hard for them to identify the early development of such ulcer.

DFU is a serious health issue that drastically affects diabetes patients' life, and burdens healthcare system. DFU with infection and ischaemia can significantly prolong treatment. It can lead to limb amputation, and even result in patients' death [18]. Furthermore, this put significant pressure onto the healthcare systems, in both time required to treat patients and cost of treatment [4].

To avoid above complications and financial burden, developing an early detection and regular monitoring is crucial, specially for high risk patients. Recently, detecting DFU from images has shown promising results [19]; in this paper we employ the concept of semi-supervised learning to address the classification of DFU patches [6,11,18] into one of four classes:

- Infection
- Ischaemia
- Both (Infection and Ischaemia)
- None

Deep Learning based methods have achieved immense performance in variety of computer vision related tasks [9–11,19]. However, their strong performance highly depends on the number of available images included in the provided training set. It is evident that there is not always possible to gather sufficiently large data set that can result in high performance deep learning classifier. This issue is more noticeable, when working with medical data that require professional labelling, consequently demand high labor cost and time. To deal with the cost limitation and lack of data, and in addition to taking advantage of the well practiced concept of transfer learning, it is common to employ data augmentation which usually consist of random rotation, zoom and other techniques of processing on the images. The problem with this common practice is that can lead to over-fitting, which would result in poor generalisation. In this paper, we aim to address both above limitations. First, we include unlabeled data in training through employing the concept of semi-supervised learning [13,15,17].

While earlier work limited the use of unlabeled data to pre-training stage [5,12], more recent researches studied exploiting unlabeled data in the entire training stage [5,12].

Recent advancements in semi-supervised deep learning methods are commonly gained by modifying the loss function, which can be achieved by adding regularization over the unlabelled data [16]. However, as new unlabeled data becomes available, all these methods must retrain the deep network in order to take advantage of such newly available data. That is both time and resource consuming. In addition, the majority of the recent approaches aim to fuse the inputs into coherent clusters by adding noise and smoothing the mapping function locally. In cases when the raw data already contain noise, such approaches would no longer be relevant.

In addition, instead of employing augmentation, and to avoid the risk of overfitting, we generate a rotation and zoom tolerant linear subspace for each image (Fig. 1).

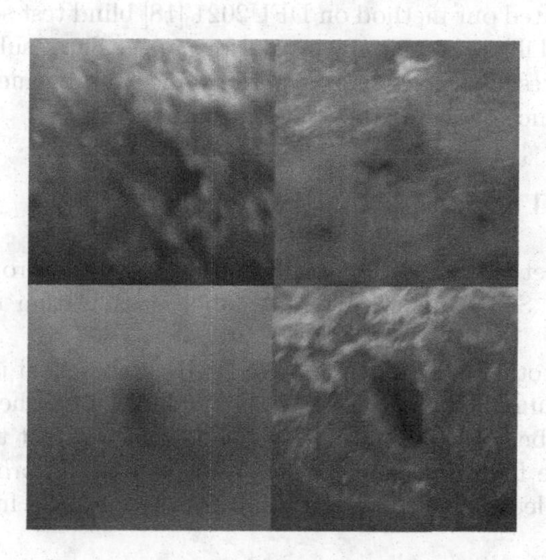

Fig. 1. Examples of DFU patches [18].

Specifically, in this work, we propose a novel 3-stage deep linear subspace analysis, for semi-supervised classification of DFU patches into Infection, Ischaemia, Both and None. As such, we start by employing the concept of transfer learning, and generate discriminate set of visual features through extracting the output of a mid layer of a pre-trained Xception [8] network. Then, to create a shift invariant representing of any given image, we use augmentation tools to create an image set for each given image. Next, we map such image sets into linear subspaces which are points on Grassmann manifold topological space. That is then followed by employing the concept of relational divergence [2] to analyse the images. That would generate the final representor of each image as a vector of relationships, and enables the method to take advantage of unlabeled data. Finally, we use the computed features in stage 2 to train Random Forest for classification purpose.

The **contributions** of this work are summaries bellow:

- The proposed method offers a novel pipeline for deep subspace analysis for semi-supervised multi-label classification (Infection, and Ischaemia) of DFU patches.
- The proposed pipeline generates an shift invariant representor of an image by first generating an image set from each given image through employing

augmentation tools. It then maps each image set to a linear subspace which is a point on a Grassmann Manifold. Then, it employs the geodesic distance to redefine each point as a vector of distance to number of unlabeled images.
- The proposed method is dose not demand retraining of the deep Xception architect when more unlabeled data become available; hence is not time and resource consuming.

We have evaluated our method on DFU2021 [18] blind test set, and the result in live leader board illustrates promising performance. The result proved that the proposed method result in considerable performance improvement in comparison with the performance of the modified Xception network.

2 DFUC2021

DFUC2021 data set [18] is used to train and evaluate the proposed Deep Subspace analysis for Semi-Supervised multi-label classification of Diabetic Foot Ulcer.

DFUC2021 in total has 15,683 images of DFU patches, that is consist of: 5,955 labeled and 3,994 unlabelled DFU patches. The labeled patches can have either of the following labels: Infection, and Ischaemia (it is evident that classification would also include following labels: both, and none). The proposed method is evaluated via live leader board on the blind test set of 5,734 images.

3 Methodology

The proposed Deep subspace analysis for semi-supervised multi-label classification is constructed of 3 main stages, detailed in bellow sections. Figure 2 Demonstrates the summary of the proposed method.

3.1 Stage 1: Transfer Learning

Number of recent semi-supervised feature extraction methods focused on finding latent representations of the input data using deep neural network.

Unlike those methods, we first employ the concept of transfer learning, on the labeled data only, to generate discriminative features.

That separates the features extraction technique from the rest of the classification stages, and results in massive time and cost reduction for retraining the method, when more unlabeled data become available.

We modify a pre-train Xception network with imagenet weights (Fig. 3), in the proposed pipeline. Extreme Inception (Xception) [8] is a deep convolutional neural network based architecture that have gained its success through decoupling the mapping of cross-channels correlations and spatial correlations in the feature maps of convolutional neural networks (Fig. 2). In other words, Xception network is a linear stack of depth-wise separable convolution layers with residual connections. The Xception network architect have following characteristics:

- That contains 36 convolutional layers.
- The 36 convolutional layers are structured into 14 modules.
- All the modules, except the first and the last, have linear residual connections around them.

Specifically, in stage 1, we modify Xception network by removing the last layer and adding two fully connected layers of size 128 and 2 to re-train the network for multi-label classification.

To address the imbalanced data, we increase the penalty weight for Ischaemia class compared to the one for Infection class.

Then, following the common practice, we train the modified Xception network in two steps; when first we freeze the original layers and train the final two layers. Then, we train the entire network with a lower learning rate. Finally, the output of a mid-layer is extracted to represent the descriptive features of each given image.

3.2 Stage 2: Deep Subspace-Based Descriptors

In this stage, to ensure that each image representation is shift invariant we employ image augmentation tools and generate an image set for each image. Then we map the image sets into points on Grassmann manifold by generating a linear subspace representation for each image. Finally, in order to take advantage of the unlabeled data, we use geodesic distance to compute the distance between labeled and number of unlabeled data; then, represent each image as a vector of relations.

In this section, we first explain the Grassmann Manifold topological space and geodesic distance that is used to calculate distance between point on Grassmann Manifold. Then we explain the transformation of an image into deep subspace-based relational vector.

Grassmann Manifold. In this study we are interested in a type of Riemannian manifolds, namely the Grassmann manifolds. Manifolds are smooth, curved surfaces embedded in higher dimensional Euclidean spaces and formally defined as follows [1]:

Definition 1. *A topological space $\mathcal{M}$ is called a manifold if:*

- *$\mathcal{M}$ is Hausdorff*[1]*, i.e. every pair $\mathbf{X}$, $\mathbf{Y}$ can be separated by two disjoint open sets.*
- *$\mathbf{M}$ is locally Euclidean, that is, for every $\mathbf{X} \in \mathbf{M}$ there exists an open set $U \subset \mathbf{M}$ with $\mathbf{X} \in \mathbf{U}$ and an open set $V \subset \mathbf{R}^n$ with a homeomorphism $\varphi : U \to V$.*

[1] In a Hausdorff space, distinct points have disjoint neighbourhoods. This property is important to establish the notion of a differential manifold, as it guarantees that convergent sequences have a single limit point.

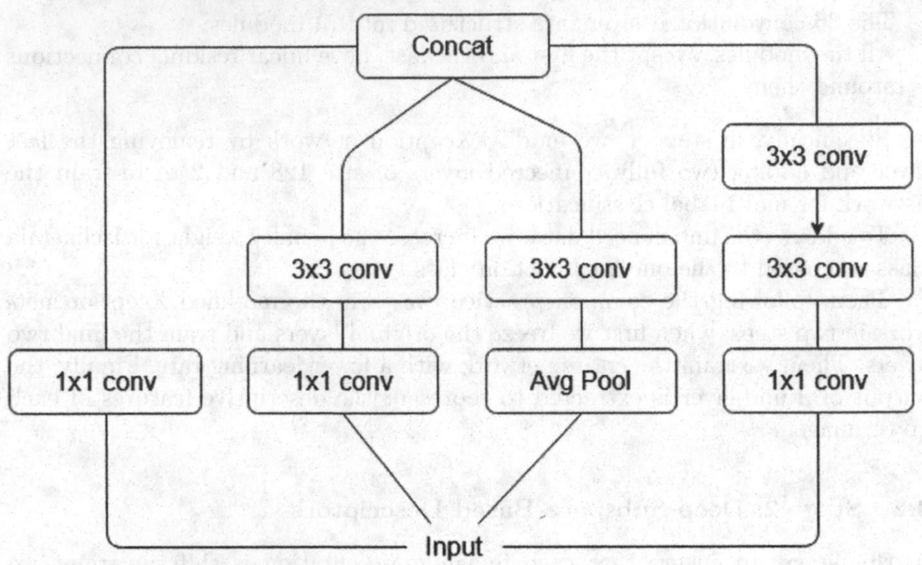

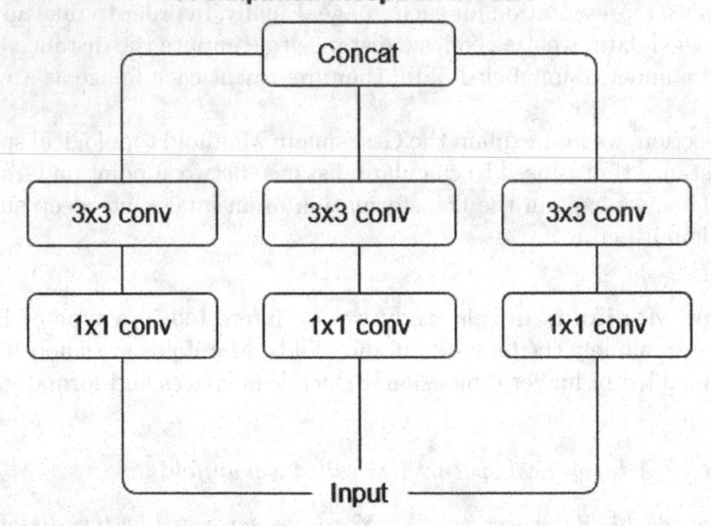

Fig. 2. The above image provides the detail of the Xception network employed as baseline [8]

Layer (type)	Output Shape	Param #
input_4 (InputLayer)	[(None, 150, 150, 3)]	0
sequential (Sequential)	(None, 150, 150, 3)	0
xception (Functional)	(None, 5, 5, 2048)	20861480
global_average_pooling2d_1 (	(None, 2048)	0
dense_2 (Dense)	(None, 128)	262272
dense_3 (Dense)	(None, 2)	258

Total params: 21,124,010
Trainable params: 21,069,482
Non-trainable params: 54,528

Fig. 3. The above image provides the detail of the modified version of Xception network employed in the proposed pipeline

To formally define a Grassmann manifold and its geometry, we need to define the quotient space of a manifold. A quotient space of a manifold, can be defined as the result of "putting together" certain points of the manifold. Formally, given $\sim_\psi$ as an equivalence relation on $\mathbf{M}$, the quotient space $\Upsilon = \mathbf{M}/\sim_\psi$ is defined to be the set of equivalence classes of elements of $\mathbf{M}$, i.e.

$$\Upsilon = \{[\mathbf{X}] : \mathbf{X} \in \mathbf{M}\} = \{[\mathbf{Y} \in \mathbf{M} : \mathbf{Y} \sim_\psi \mathbf{X}] : \mathbf{X} \in \mathbf{M}\}.$$

Definition 2. *Specifically, a Grassmann manifold is a quotient space of the special orthogonal group*[2] *$SO(n)$ and is defined as a set of p-dimensional linear subspaces of $\mathbf{R}^n$ [1].*

In practice an element $\mathbf{X}$ of $\mathbf{G}n, p$ is represented by an orthonormal basis as a $n \times p$ matrix, i.e., $\mathbf{X}^T\mathbf{X} = \mathbf{I}_p$. The geodesic distance between two points on the Grassmann manifold can be computed as:

$$d_G(\mathbf{X}, \mathbf{Y}) = \|\Theta\|_2 \tag{1}$$

where $\Theta = [\theta_1, \theta_2, \cdots, \theta_p]$ is the principal angle vector, i.e.:

$$\cos(\theta_i) = \max_{\vec{x}_i \in \mathbf{X},\, \vec{y}_j \in \mathbf{Y}} \vec{x}_i^T \vec{y}_j \tag{2}$$

subject to $\mathbf{x}_i^T\mathbf{x}_i = \mathbf{y}_i^T\mathbf{y}_i = 1$, $\mathbf{x}_i^T\mathbf{x}_j = \mathbf{y}_i^T\mathbf{y}_j = 0$, $i \neq j$. The principal angles have the property of $\theta_i \in [0, \pi/2]$ and can be computed through SVD of $\mathbf{X}^T\mathbf{Y}$.

[2] Special orthogonal group $SO(n)$ is the space of all $n \times n$ orthogonal matrices with the determinant +1. It is not a vector space but a differentiable manifold, i.e., it can be locally approximated by subsets of a Euclidean space.

Geodesic-Based Relational Representation. First, borrowing the concept of relational divergence [2], we employ K-medians clustering on unlabeled data to generate their representatives. Then we generate linear subspace for each labeled data and the centroid images of unlabeled data.

Next, to generate a linear subspace representative for each image, we start by employing data augmentation to generate image set representation of each image. Then, we represent each image as an output of the mid layer of modified Xception network; which is then followed by calculating the strongest Eigen vectors through computing Singular Value Decomposition (SVD) of the deep image-set. That maps each image set into a linear subspace which is a point on Grassmann Manifold.

Then, we calculate the geodesic distance between each labeled image to the K-medians centers of unlabeled images. That is constructed by calculating the distance between extracted features computed in stage 1.

Let $\mathbf{\check{L}}$ represent the labeled training data, and $\mathbf{\check{U}}$ represent the Un-labeled training data:

$$\mathbf{\check{L}} = [\mathbf{l_1}, \mathbf{l_2}, ...\mathbf{l_m}]$$
$$\mathbf{l_i} = [n_{i,1}, n_{i,2}, ..., n_{i,128}]$$

$$\mathbf{\check{U}} = [\mathbf{u_1}, \mathbf{u_2}, ...\mathbf{u_p}]$$
$$\mathbf{u_j} = [n_{j,1}, n_{j,2}, ..., n_{j,128}],$$

where vector $\mathbf{l_i}$ and $\mathbf{u_j}$ are the deep representation of a labeled image I_i, and un-labeled image U_j respectively. That is the feature vector extracted from the mid-layer of the modified Xception architect.

After performing Kmedians on un-labeled training data, the un-labeled data would be represented as matrix $\mathbf{\check{C}}$:

$$\mathbf{\check{C}} = [\mathbf{c_1}, \mathbf{c_2}, ...\mathbf{c_p}]$$
$$\mathbf{c_j} = [n_{j,1}, n_{j,2}, ..., n_{j,\alpha}],$$

where vector $\mathbf{c_k}$ is the deep representation of a centroid image $\mathbf{u_k}$. For this work we have used $\alpha = 200$, that means the Kmedians would compute the index of 200 centroids for un-labeled training data.

At this point, we transfer each image into an image set using augmentation techniques:

$$\forall \mathbf{L_j} \in \mathbf{\check{L}} : \mathbf{L_j} = [\mathbf{l_{j1}}, \mathbf{l_{j2}}, ...\mathbf{l_{jp}}] \quad \text{where} \quad \mathbf{l_{jf}} = [n_{jf,1}, n_{jf,2}, ..., n_{jf,\alpha}]$$
$$\mathbf{\dot{L}_j} = \left[\mathbf{\bar{u}_{l_{j1}}}, \mathbf{\bar{u}_{l_{j2}}}, ..., \mathbf{\bar{u}_{l_{jp}}}\right] \quad \text{where } \mathrm{SVD}(\mathbf{L_j}) = \mathbf{\hat{U}_{L_j}} \mathbf{\hat{\Xi}_{L_j}} \mathbf{\hat{V}_{L_j}}, \text{ and}$$
$$\forall \mathbf{C_j} \in \mathbf{\check{C}} : \mathbf{C_j} = [\mathbf{c_{j1}}, \mathbf{c_{j2}}, ...\mathbf{c_{jm}}] \quad \text{where} \quad \mathbf{c_{jf}} = [n_{jf,1}, n_{jf,2}, ..., n_{jf,\alpha}]$$
$$\mathbf{\dot{C}_j} = \left[\mathbf{\bar{u}_{c_{j1}}}, \mathbf{\bar{u}_{c_{j2}}}, ..., \mathbf{\bar{u}_{c_{jp}}}\right] \quad \text{where } \mathrm{SVD}(\mathbf{C_j}) = \mathbf{\hat{U}_{C_j}} \mathbf{\hat{\Xi}_{C_j}} \mathbf{\hat{V}_{C_j}},$$

where $\mathbf{\dot{L}_j}$ and $\mathbf{\dot{C}_j}$ are each a linear subspace representing labeled image L_j, and a centroid of unlabeled data C_j respectively.

Finally, we represent each Labeled image $\mathbf{l_i}$ through calculating the geodesic distance between $\dot{\mathbf{L_i}}$ and all the $\dot{\mathbf{C_j}} \in \dot{\check{\mathbf{C}}}$.

$$\text{for } \mathbf{L_j} \in \dot{\check{\mathbf{L}}} \text{ and, } \mathbf{C_j} \in \dot{\check{\mathbf{C}}}:$$

$$\|\mathbf{d_i}\| = \left(G_d(\dot{\mathbf{L_j}}, \dot{\mathbf{C_1}}), G_d(\dot{\mathbf{L_j}}, \dot{\mathbf{C_2}}), ..., G_d(\dot{\mathbf{L_j}}, \dot{\mathbf{C_\alpha}})\right),$$

where each image $\mathbf{l_i}$ is now represented as $\|\mathbf{d_{Gi}}\|$ detailed in the Eq. 1 (Fig. 4).

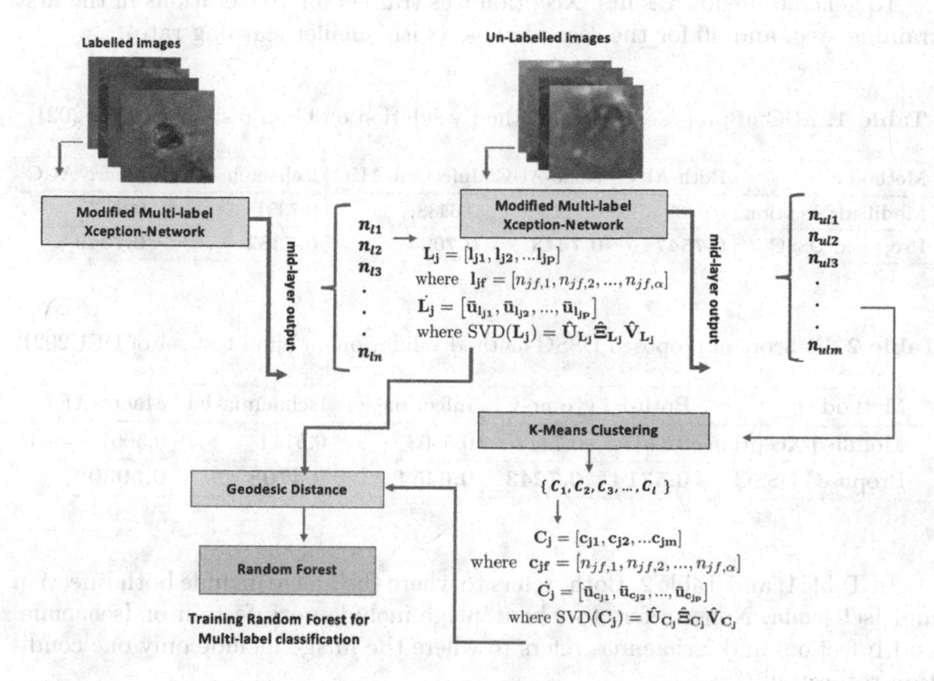

Fig. 4. The above image provides a summary of the proposed DSSC pipeline.

3.3 Stage 3: Final Classification

Finally, we employ Multi-Label Random Forest (MLRF) [14] classification method on the resulting feature vector. MLRF, is a multi-label classification method based on a variation of random forest. It uses a new label set partition method to transform multi-label data sets into multiple single-label data sets. That can optimize the label subset partition, by discovering the correlated labels. That employs an on-line kNNs-like sampling method for each generated single-label subset ignorer to learn a random forest classifier.

4 Results and Discussion

To evaluate the proposed method, we have tested its performance on blind test set of Diabetic Foot Ulcers dataset (DFUC2021) [18]. We have used 5,955 DFU images for training, 5,734 for blind testing [18]. The ground truth labels comprise of four classes: control, infection, ischaemia and both conditions. The results indicate a considerable improvement when using the proposed semi-supervised method compared to solely relying on transfer learning through using the modified version of Xception. Bellow table summarises our finding.

It is important to note that more complex modification of Xception with more data augmentation would result in the better performance for both Xception and DSSC. That would be developed and evaluated in future work.

To generate bellow results, Xception was trained for 10 iterations in the first training step, and 40 for the second stage (with smaller learning rate).

Table 1. AUC of proposed DSSC method validation on blind test set of DFU2021.

Method	Both-AUC	None-AUC	Infection-AUC	Ischaemia-AUC	Macro-AUC
Modified Xception	0.6483	0.7215	0.6438	0.7331	0.6867
Proposed DSSC	**0.7547**	**0.7443**	**0.7024**	**0.7382**	**0.7349**

Table 2. F1-Score of proposed DSSC method validation on blind test set of DFU2021.

Method	Both-F1	None-F1	Infection-F1	Ischaemia-F1	Macro-AF1
Modified Xception	0.3737	0.7112	0.5503	0.5111	0.5067
Proposed DSSC	**0.5314**	**0.7243**	**0.6454**	**0.4708**	**0.5930**

In Table 1, and Table 2: Both- refers to where the image include both Infection and Ischaemia, None- refers to where image include no Infection or Ischaemia, and Infection- and Ischaemia- refers to where the image include only one condition respectfully.

The above results indicate that the proposed DSSC method results in considerable improvement in performance compared to Xception, through taking advantage of unlabeled data, while ensuring the ease of retraining for new unlabeled data (when become available).

5 Conclusion

In this work, we propose a novel deep subspace analysis method for semi-supervised multi-label classification (DSSCC) of DFU images. The proposed method have two main differences compared to the recent state of the art deep semi-supervised methods. First, unlike recent research works in deep semi-supervised methods, the proposed pipeline dose not augment data during training; instead, to generate a shift invariant representative, it transfer each image into a linear subspace, and analysis them using Grassmann manifold geometry. Moreover, the method considered that the likelihood of new unlabelled data becoming available is always higher compared to that for labeled data. Thus, the proposed method is designed so the retraining of network with more unlabeled data would not be time and resource consuming. The evaluation of DSSC on blind test set of DFU2021 shows considerable improvement compared to the performance of solely relying on labeled data using Xception. That proves the efficiency of the proposed Deep Relation-based Semi-Supervised.

Acknowledgement. This work was supported by the RMIT Alumni and Philanthropy and the Malcolm Moore Industry Research Award, and by Professor David Ascher's Lab at Baker Heart and Diabetes Institute.

References

1. Alavi, A.: Image analysis on symmetric positive definite manifolds (2014)
2. Alavi, A., Harandi, M.T., Sanderson, C.: Relational divergence based classification on Riemannian manifolds. In: 2013 IEEE Workshop on Applications of Computer Vision (WACV), pp. 111–116. IEEE (2013)
3. Armstrong, D.G., Boulton, A.J., Bus, S.A.: Diabetic foot ulcers and their recurrence. N. Engl. J. Med. **376**(24), 2367–2375 (2017)
4. Association, A.D., et al.: Economic costs of diabetes in the US in 2017. Diabetes Care **41**(5), 917–928 (2018)
5. Bennett, K., Demiriz, A., et al.: Semi-supervised support vector machines. In: Advances in Neural Information Processing Systems, pp. 368–374 (1999)
6. Cassidy, B., et al.: The DFUC 2020 dataset: analysis towards diabetic foot ulcer detection. touchREVIEWS Endocrinol. **17**, 5–11 (2021). https://doi.org/10.17925/EE.2021.17.1.5
7. Cho, N., et al.: IDF diabetes atlas: global estimates of diabetes prevalence for 2017 and projections for 2045. Diabetes Res. Clin. Pract. **138**, 271–281 (2018)
8. Chollet, F.: Xception: deep learning with depthwise separable convolutions. In: Proceedings of the IEEE Conference on Computer Vision and Pattern Recognition, pp. 1251–1258 (2017)
9. Goyal, M., Reeves, N.D., Davison, A.K., Rajbhandari, S., Spragg, J., Yap, M.H.: DFUNet: convolutional neural networks for diabetic foot ulcer classification. IEEE Trans. Emerg. Top. Comput. Intell. **4**(5), 728–739 (2018). https://doi.org/10.1109/TETCI.2018.2866254
10. Goyal, M., Reeves, N.D., Rajbhandari, S., Yap, M.H.: Robust methods for real-time diabetic foot ulcer detection and localization on mobile devices. IEEE J. Biomed. Health Inform. **23**(4), 1730–1741 (2019). https://doi.org/10.1109/JBHI.2018.2868656
11. Goyal, M., Reeves, N.D., Rajbhandari, S., Ahmad, N., Wang, C., Yap, M.H.: Recognition of ischaemia and infection in diabetic foot ulcers: dataset and techniques. Comput. Biol. Med. **117**, 103616 (2020). https://doi.org/10.1016/j.compbiomed.2020.103616. http://www.sciencedirect.com/science/article/pii/S0010482520300160
12. Joachims, T., et al.: Transductive inference for text classification using support vector machines. In: ICML, vol. 99, pp. 200–209 (1999)
13. Khaki, S., Pham, H., Han, Y., Kuhl, A., Kent, W., Wang, L.: DeepCorn: a semi-supervised deep learning method for high-throughput image-based corn kernel counting and yield estimation. Knowl.-Based Syst. **218**, 106874 (2021)

14. Liu, F., Zhang, X., Ye, Y., Zhao, Y., Li, Y.: MLRF: multi-label classification through random forest with label-set partition. In: Huang, D.-S., Han, K. (eds.) ICIC 2015. LNCS (LNAI), vol. 9227, pp. 407–418. Springer, Cham (2015). https://doi.org/10.1007/978-3-319-22053-6_44
15. Shrivastava, A., Pillai, J.K., Patel, V.M., Chellappa, R.: Learning discriminative dictionaries with partially labeled data. In: 2012 19th IEEE International Conference on Image Processing, pp. 3113–3116. IEEE (2012)
16. van Engelen, J.E., Hoos, H.H.: A survey on semi-supervised learning. Mach. Learn. **109**(2), 373–440 (2019). https://doi.org/10.1007/s10994-019-05855-6
17. Wu, H., Prasad, S.: Semi-supervised deep learning using pseudo labels for hyperspectral image classification. IEEE Trans. Image Process. **27**(3), 1259–1270 (2017)
18. Yap, M.H., Cassidy, B., Pappachan, J.M., O'Shea, C., Gillespie, D., Reeves, N.D.: Analysis towards classification of infection and ischaemia of diabetic foot ulcers. In: Proceedings of the IEEE EMBS International Conference on Biomedical and Health Informatics (BHI 2021), pp. 1–4 (2021). https://doi.org/10.1109/BHI50953.2021.9508563
19. Yap, M.H., et al.: Deep learning in diabetic foot ulcers detection: a comprehensive evaluation. Comput. Biol. Med. **135**, 104596 (2021)

Author Index

Printed in the United States
by Baker & Taylor Publisher Services